I0508912

Marx Joyce
Defoe Hardy Emerson Cooper Austen
Abbott Machiavelli Hugo
Melville Montaigne Chesterton Eliot
Stoker Carroll Haggard Molière Grimm
Christie Maupassant Byron
Wilde Fitzgerald Engels Schiller
Garnett Einstein Hawthorne Smith Kafka
Goethe Kipling Doyle Hall
Baum Cotton Dostoyevsky Willis
Leslie Dumas Henry Nietzsche
Flaubert Turgenev Balzac
Stockton Vatsyayana Crane
Burroughs Verne
Curtis Tocqueville Whitman Gogol Vinci
Homer Widger Tolstoy Busch
Darwin Thoreau Twain
Potter Freud Zola Plato Scott
Kant Jowett Lawrence Harte
Andersen Dickens Hesse
London Descartes Burton
Poe Aristotle Wells Voltaire
Hale James Hastings Cervantes Cooke
Bunner Shakespeare Irving
Richter Chambers
Doré da Benedict Alcott
Dante Shaw Wodehouse
Swift Chekhov Pushkin
Newton

tredition

tredition was established in 2006 by Sandra Latusseck and Soenke Schulz. Based in Hamburg, Germany, tredition offers publishing solutions to authors and publishing houses, combined with worldwide distribution of printed and digital book content. tredition is uniquely positioned to enable authors and publishing houses to create books on their own terms and without conventional manufacturing risks.

For more information please visit: www.tredition.com

TREDITION CLASSICS

This book is part of the TREDITION CLASSICS series. The creators of this series are united by passion for literature and driven by the intention of making all public domain books available in printed format again - worldwide. Most TREDITION CLASSICS titles have been out of print and off the bookstore shelves for decades. At tredition we believe that a great book never goes out of style and that its value is eternal. Several mostly non-profit literature projects provide content to tredition. To support their good work, tredition donates a portion of the proceeds from each sold copy. As a reader of a TREDITION CLASSICS book, you support our mission to save many of the amazing works of world literature from oblivion. See all available books at www.tredition.com.

 ## Project Gutenberg

The content for this book has been graciously provided by Project Gutenberg. Project Gutenberg is a non-profit organization founded by Michael Hart in 1971 at the University of Illinois. The mission of Project Gutenberg is simple: To encourage the creation and distribution of eBooks. Project Gutenberg is the first and largest collection of public domain eBooks.

The Minds and Manners of Wild Animals A Book of Personal Observations

William Temple Hornaday

Imprint

This book is part of TREDITION CLASSICS

Author: William Temple Hornaday
Cover design: Buchgut, Berlin – Germany

Publisher: tredition GmbH, Hamburg - Germany
ISBN: 978-3-8424-6079-9

www.tredition.com
www.tredition.de

Copyright:
The content of this book is sourced from the public domain.

The intention of the TREDITION CLASSICS series is to make world literature in the public domain available in printed format. Literary enthusiasts and organizations, such as Project Gutenberg, worldwide have scanned and digitally edited the original texts. tredition has subsequently formatted and redesigned the content into a modern reading layout. Therefore, we cannot guarantee the exact reproduction of the original format of a particular historic edition. Please also note that no modifications have been made to the spelling, therefore it may differ from the orthography used today.

CONTENTS

I. A SURVEY OF THE FIELD

I. THE LAY OF THE LAND
II. WILD ANIMAL TEMPERAMENT & INDIVIDUALITY
III. THE LANGUAGE OF ANIMALS
IV. THE MOST INTELLIGENT ANIMALS
V. THE RIGHTS OF WILD ANIMALS

II. MENTAL TRAITS OF WILD ANIMALS

VI. THE BRIGHTEST MINDS AMONG ANIMALS
VII. KEEN BIRDS AND DULL MEN
VIII. THE MENTAL STATUS OF THE ORANG-UTAN
IX. THE MAN-LIKENESS OF THE CHIMPANZEE
X. THE TRUE MENTAL STATUS OF THE GORILLA
XI. THE MIND OF THE ELEPHANT
XII. THE MENTAL AND MORAL TRAITS OF BEARS
XIII. MENTAL TRAITS OF A FEW RUMINANTS
XIV. MENTAL TRAITS OF A FEW RODENTS
XV. THE MENTAL TRAITS OF BIRDS
XVI. THE WISDOM OF THE SERPENT
XVII. THE TRAINING OF WILD ANIMALS

III. THE HIGHER PASSIONS

XVIII. THE MORALS OF WILD ANIMALS
XIX. THE LAWS OF THE FLOCKS AND HERDS
XX. PLAYS AND PASTIMES OF WILD ANIMALS
XXI. COURAGE IN WILD ANIMALS

IV. THE BASER PASSIONS

XXII. FEAR AS A RULING PASSION
XXIII. FIGHTING AMONG WILD ANIMALS
XXIV. WILD ANIMAL CRIMINALS AND CRIME
XXV. FIGHTING WITH WILD ANIMALS

THE CURTAIN.

PREFACE

During these days of ceaseless conflict, anxiety and unrest among men, when at times it begins to look as if "the Caucasian" really is "played out," perhaps the English-reading world will turn with a sigh of relief to the contemplation of wild animals. At all events, the author has found this diversion in his favorite field mentally agreeable and refreshing.

In comparison with some of the alleged men who now are cursing this earth by their baneful presence, the so-called "lower animals" do not seem so very "low" after all! As a friend of the animals, this is a very proper time in which to compare them with men. Furthermore, if thinking men and women desire to know the leading facts concerning the intelligence of wild animals, it will be well to consider them now, before the bravest and the best of the wild creatures of the earth go down and out under the merciless and inexorable steam roller that we call Civilization.

The intelligence and the ways of wild animals are large subjects. Concerning them I do not offer this volume as an all-in-all production. Out of the great mass of interesting things that might have been included, I have endeavored to select and set forth only enough to make a good series of sample exhibits, without involving the general reader in a hopelessly large collection of details. The most serious question has been: What shall be left out?

Mr. A. R. Spofford, first Librarian of Congress, used to declare that "Books are made from books"; but I call the reader to bear witness that this volume is not a mass of quotations. A quoted authority often can be disputed, and for this reason the author has found considerable satisfaction in relying chiefly upon his own testimony.

Because I always desire to know the *opinions* of men who are writing upon their own observations, I have felt free to express my own conclusions regarding the many phases of animal intelligence as their manifestation has impressed me in close-up observations.

I have purposely avoided all temptations to discuss the minds and manners of domestic animals, partly because that is by itself a large subject, and partly because their minds have been so greatly influenced by long and close association with man. The domestic mammals and birds deserve independent treatment.

A great many stories of occurrences have been written into this volume, for the purpose of giving the reader all the facts in order that he may form his own opinions of the animal mentality displayed.

Most sincerely do I wish that the boys and girls of America, and of the whole world, may be induced to believe that *the most interesting thing about a wild animal is its mind and its reasoning,* and that a dead animal is only a poor decaying thing. If the feet of the young men would run more to seeing and studying the wild creatures and less to the killing of them, some of the world's valuable species might escape being swept away tomorrow, or the day after.

The author gratefully acknowledges his indebtedness to Munsey's Magazine, McClure's Magazine and the Sunday Magazine Syndicate for permission to copy herein various portions of his chapters from those publications.

W. T. H.

The Anchorage, Stamford, Conn. December 19, 1921.

ILLUSTRATIONS

Overpowering Curiosity of a Mountain Sheep
Christmas at the Primates' House
The Trap-Door Spider's Door and Burrow
Hanging Nest of the Baltimore Oriole Great
Hanging Nests of the Crested Cacique
"Rajah," the Actor Orang-Utan
Thumb-Print of an Orang-Utan
The Lever That Our Orang-Utan Invented
Portrait of a High-Caste Chimpanzee
The Gorilla With the Wonderful Mind
Tame Elephants Assisting in Tying a Wild Captive
Wild Bears Quickly Recognize
Protection Alaskan Brown Bear,
"Ivan," Begging for Food
The Mystery of Death
The Steady-Nerved and Courageous Mountain Goat
Fortress of an Arizona Pack-Rat
Wild Chipmunks Respond to Man's Protection
An Opossum Feigning Death
Migration of the Golden Plover. (Map)
Remarkable Village Nests of the Sociable Weaver Bird
Spotted Bower-Bird, at Work on Its Unfinished Bower Hawk-Proof
Nest of a Cactus Wren
A Peace Conference With an Arizona Rattlesnake
Work Elephant Dragging a Hewn Timber The Wrestling Bear,
"Christian," and His Partner
Adult Bears at Play
Primitive Penguins on the Antarctic Continent, Unafraid of Man
Richard W. Rock and His Buffalo Murderer
"Black Beauty" Murdering "Apache"

THE MINDS AND MANNERS OF WILD ANIMALS

MAN AND THE WILD ANIMALS

If every man devoted to his affairs, and to the affairs of his city and state, the same measure of intelligence and honest industry that every warm-blooded wild animal devotes to its affairs, the people of this world would abound in good health, prosperity, peace and happiness.

To assume that every wild beast and bird is a sacred creature, peacefully dwelling in an earthly paradise, is a mistake. They have their wisdom and their folly, their joys and their sorrows, their trials and tribulations.

As the alleged lord of creation, it is man's duty to know the wild animals truly as they are, in order to enjoy them to the utmost, to utilize them sensibly and fairly, and to give them a square deal.

I. A SURVEY OF THE FIELD

I

THE LAY OF THE LAND

There is a vast field of fascinating human interest, lying only just outside our doors, which as yet has been but little explored. It is the Field of Animal Intelligence.

Of all the kinds of interest attaching to the study of the world's wild animals, there are none that surpass the study of their minds, their morals, and the acts that they perform as the results of their mental processes.

In these pages, the term "animal" is not used in its most common and most restricted sense. It is intended to apply not only to quadrupeds, but also to all the vertebrate forms, — mammals, birds, reptiles, amphibians and fishes.

For observation and study, the whole vast world of living creatures is ours, throughout all zones and all lands. It is not ours to flout, to abuse, or to exterminate as we please. While for practical reasons we do not here address ourselves to the invertebrates, nor even to the sea-rovers, we can not keep them out of the background of our thoughts. The living world is so vast and so varied, so beautiful and so ugly, so delightful and so terrible, so interesting and so commonplace, that each step we make through it reveals things different and previously unknown.

The Frame of Mind. To the inquirer who enters the field of animal thought with an open mind, and free from the trammels of egotism and fear regarding man's place in nature, this study will prove an endless succession of surprises and delights. In behalf of the utmost tale of results, the inquirer should summon to his aid his rules of evidence, his common sense, his love of fair play, and the inexorable logic of his youthful geometry.

And now let us clear away a few weeds from the entrance to our field, and reveal its cornerstones and boundary lines. To a correct understanding of any subject a correct point of view is absolutely essential.

In a commonplace and desultory way man has been mildly interested in the intelligence of animals for at least 30,000 years. The Cro-Magnons of that far time possessed real artistic talent, and on the smooth stone walls and ceilings of the caves of France they drew many wonderful pictures of mammoths, European bison, wild cattle, rhinoceroses and other animals of their period. Ever since man took unto himself certain tractable wild animals, and made perpetual thralls of the horse, the dog, the cat, the cattle, sheep, goats and swine, he has noted their intelligent ways. Ever since the first caveman began to hunt wild beasts and slay them with clubs and stones, the two warring forces have been interested in each other, but for about 25,000 years I think that the wild beasts knew about as much of man's intelligence as men knew of theirs.

I leave to those who are interested in history the task of revealing the date, or the period, when scholarly men first began to pay serious attention to the animal mind.

In 1895 when Mr. George J. Romanes, of London, published his excellent work on "Animal Intelligence," on one of its first pages he blithely brushed aside as of little account all the observations, articles and papers on his subject that had been published previous to that time. Now mark how swiftly history can repeat itself, and also bring retribution.

In 1910 there arose in the United States of America a group of professional college-and-university animal psychologists who set up the study of "animal behavior." They did this so seriously, and so determinedly, that one of the first acts of two of them consisted in joyously brushing aside as of no account whatever, and quite beneath serious consideration, everything that had been seen, done and said previous to the rise of their group, and the laboratory Problem Box. In view of what this group has accomplished since 1910, with their "problem boxes," their "mazes" and their millions of "trials by error," expressed in solid pages of figures, the world of animal lovers is entitled to smile tolerantly upon the cheerful assumptions of ten years ago.

But let it not at any time be assumed that we are destitute of problem boxes; for the author has two of his own! One is called the Great Outdoors, and the other is named the New York Zoological Park. The first has been in use sixty years, the latter twenty-two years. Both are today in good working order, but the former is not quite as good as new.

A Preachment to the Student. In studying the wild-animal mind, the boundary line between Reality and Dreamland is mighty easy to cross. He who easily yields to seductive reasoning, and the call of the wild imagination, soon will become a dreamer of dreams and a seer of visions of things that never occurred. The temptation to place upon the simple acts of animals the most complex and farfetched interpretations is a trap ever ready for the feet of the unwary. It is better to see nothing than to see a lot of things that are not true.

In the study of animals, we have long insisted that *to the open eye and the thinking brain, truth is stranger than fiction.* But Truth does not always wear her heart upon her sleeve for zanies to peck at. Unfortunately there are millions of men who go through the world looking at animals, but not seeing them.

Beware of setting up for wild animals impossible mental and moral standards. The student must not deceive himself by overestimating mental values. If an estimate must be made, make it under the mark of truth rather than above it. While avoiding the folly of idealism, we also must shun the ways of the narrow mind, and the eyes that refuse to see the truth. Wild animals are not superhuman demigods of wisdom; but neither are they idiots, unable to reason from cause to effect along the simple lines that vitally affect their existence.

Brain-owning wild animals are not mere machines of flesh and blood, set agoing by the accident of birth, and running for life on the narrow-gauge railway of Heredity. They are not "Machines in Fur and Feathers," as one naturalist once tried to make the world believe them to be. Some animals have more intelligence than some men; and some have far better morals.

What Constitutes Evidence. The best evidence regarding the ways of wild animals is one's own eye-witness testimony. Not all secondhand observations are entirely accurate. Many persons do not know how to observe; and at times some are deceived by their own eyes or ears. It is a sad fact that both those organs are easily deceived. The student who is in doubt regarding the composition of evidence will do well to spend a few days in court listening to the trial of an important and hotly contested case. In collecting real evidence, all is not gold that glitters.

Many a mind misinterprets the thing seen, sometimes innocently, and again wantonly. The nature fakir is always on the alert to see wonderful phenomena in wild life, about which to write; and by preference he places the most strained and marvellous interpretation upon the animal act. Beware of the man who always sees marvellous things in animals, for he is a dangerous guide. There is one man who claims to have seen in his few days in the woods more wonders than all the older American naturalists and sportsmen have seen added together.

Now, Nature does not assemble all her wonderful phenomena and hold them in leash to be turned loose precisely when the great Observer of Wonders spends his day in the woods. Wise men always suspect the man who sees too many marvelous things.

The Relative Value of Witnesses. It is due that a word should be said regarding "expert testimony" in the case of the wild animal. Some dust has been raised in this field by men posing as authorities on wild animal psychology, whose observations of the world's wild animals have been confined to the chipmunks, squirrels, weasels, foxes, rabbits, and birds dwelling within a small circle surrounding some particular woodland house. In another class other men have devoted heavy scientific labors to laboratory observations on white rats, domestic rabbits, cats, dogs, sparrows, turtles and newts as the handpicked exponents of the intelligence of the animals of the world!

Alas! for the human sense of Proportion!

Fancy an ethnologist studying the Eskimo, the Dog-Rib Indian, the Bushman, the Aino and the Papuan, and then proceeding to write conclusively "On the Intelligence of the Human Race."

The proper place in which to study the minds, manners and morals of wild animals is in the most thickly populated haunts of the most intelligent species. The free and untrammeled animal, busily working out its own destiny unhindered by man, is the beau-ideal animal to observe and to study. Go to the plain, the wilderness, the desert and the mountain, not merely to shoot everything on foot, but to SEE *animals at home,* and there use your eyes and your fieldglass. See what *normal wild animals* do as "behavior," and then try to find out why they do it.

The next best place for study purposes is a spacious, sanitary and well-stocked zoological park, wherein are assembled great collections of the most interesting land vertebrates that can be procured, from all over the earth. There the student can observe many new traits of wild animal character, as they are brought to the surface by captivity. There will some individuals reveal the worst traits of their species. Others will reveal marvels in mentality, and teach lessons such as no man can learn from them in the open. To study temperament, there is no place like a zoo.

Even there, however, the wisest course,—as it seems to me,—is not to introduce too many appliances as aids to mental activity, but rather to see what the animal subject thinks and does *by its own initiative.* In the testing of memory and the perceptive faculties, training for performances is the best method to pursue.

The reader has a right to know that the author of this volume has enjoyed unparalleled opportunities for the observation and study of highly intelligent wild animals, both in their wild haunts and in a great vivarium; and these combined opportunities have covered a long series of years.

Before proceeding farther, it is desirable to define certain terms that frequently will be used in these pages.

THE ANIMAL BRAIN is the generator of the mind, and the clearing- house of the senses. As a mechanism, the brain of man is the most perfect, and in the descent through the mammals, birds, reptiles, amphibians and fishes, the brain progressively is simplified in form and function.

THOUGHT is the result of the various processes of the brain and nervous system, stimulated by the contributions of the senses.

SANITY is the state of normal, orderly and balanced thought, as formulated by a healthy brain.

INSANITY is a state of mental disease, resulting in disordered, unbalanced and chaotic thought, destitute of reason.

REASON is the manifestation of correct observation and healthful thought which recognizes both cause and effect, and leads from premise to conclusion. INTELLIGENCE is created by the possession of knowledge either inherited or acquired. It may be either latent or active; and it is the forerunner of reason.

INSTINCT is the knowledge or impulse which animals or men derive from their ancestors by inheritance, and which they obey, either consciously or subconsciously in working out their own preservation, increase and betterment. Instinct often functions as a sixth sense.

EDUCATION is the acquirement of knowledge by precept or by observation; but animals as well as men may be self-taught, and become self-educated, by the diligent exercise of the observing and reasoning faculties. The adjustment of a wild animal mind to conditions unknown to its ancestors is through the process of self-education, and by logical reasoning from premise to conclusion.

The wild animal must think, or die.

Animal intelligence varies in quantity and quality as much as animals vary in size. Idiots, maniacs and sleeping persons are the only classes of human beings who are devoid of intelligence and reasoning power. Idiots and maniacs also are often devoid of the common animal *instinct* that ordinarily promotes self- preservation from fire, water and high places. A heavily sleeping person is often so sodden in slumber that his senses of smell and hearing are temporarily dead; and many a sleeping man has been asphyxiated by gas or smoke, or burned to death, because his deadened senses failed to arouse him at the critical moment. (This dangerous condition of mind can be cured by efforts of the will, exercised prior to sleep, through a determination resolutely to arouse and investigate every unusual sensation that registers "danger" on any one of the senses.) The normal individual sleeps with a subconscious and sensitive mind, from which thought and reason have not been entirely eliminated.

Every act of a man or animal, vertebrate or invertebrate, is based upon either *reason* or *hereditary instinct.* It is a mistake to assume that because an organism is small it necessarily has no "mind," and none of the propelling impulse that we call thought. The largest whale may have less intelligence and constructive reasoning than a trap-door spider, a bee or an ant. To deny this is to deny the evidence of one's senses.

A MEASURE FOR ANIMAL INTELLIGENCE. The intelligence of an animal may be estimated by taking into account, separately, its mental qualities, about as follows:

1. General knowledge of surrounding conditions. 2. Powers of independent observation and reasoning. 3. Memory. 4. Comprehension under tuition. 5. Accuracy in the execution of man's orders.

Closely allied to these are the *moral qualities* which go to make up an animal's temperament and disposition, about as follows:

1. Amiability, which guarantees security to its associates. 2. Patience, or submission to discipline and training. 3. Courage, which gives self-confidence and steadiness. 4. A disposition to obedience, with cheerfulness.

All normal vertebrate animals exercise their intelligence in accordance with their own rules of logic. Had they not been able to do so, it is reasonable to suppose that they could never have developed into vertebrates, reaching even up to man himself.

According to the laws of logic, this proposition is no more open to doubt or dispute than is the existence of the Grand Canyon of the Colorado. But few persons have seen the Canyon, and far fewer ever have proven its existence by descending to its bottom; but none the less Reason admonishes all of us that the great chasm exists, and is not a debatable question.

To men and women who really know the vertebrate animals by contact with some of them upon their own levels, the reasoning power of the latter is not a debatable question. The only real question is: how far does their intelligence carry them? It is with puzzled surprise that we have noted the curious diligence of the professors of animal psychology in always writing of "animal *behavior*," and never of old-fashioned, common-sense *animal intelligence*. Can it be possible that any one of them really refuses to concede to the wild animal the possession of a mind, and a working intelligence?

Yes. Animals do reason. If any one truth has come out of all the critical or uncritical study of the animal mind that has been going on for two centuries, it is this. Animals do reason; they always have reasoned, and as long as animals live they never will cease to reason.

The higher wild animals possess and display the same fundamental passions and emotions that animate the human race. This fact is subject to intelligent analysis, discussion and development, but it is not by any means a "question" subject to debate. In the most intellectual of the quadrupeds, birds and reptiles, the display of fear, courage, love, hate, pleasure, displeasure, confidence, suspicion, jealousy, pity, greed and generosity are so plainly evident that even children can and do recognize them. To the serious and open-minded student who devotes prolonged thought to these things, they bring the wild animal very near to the "lord of creation."

To the question, "Have wild animals souls?" we reply, "That is a debatable question. Read; then think it over."

METHODS WITH THE ANIMAL MIND. In the study of animal minds, much depends upon the method employed. It seems to me that the problem- box method of the investigators of "animal behavior" leaves much to be desired. Certainly it is not calculated to develop the mental status of animals along lines of natural mental progression. To place a wild creature in a great artificial contrivance, fitted with doors, cords, levers, passages and what not, is enough to daze or frighten any timid animal out of its normal state of mind and nerves. To put a wild sapajou monkey,— weak, timid and afraid,—in a strange and formidable prison box filled with strange machinery, and call upon it to learn or to invent strange mechanical processes, is like bringing a boy of ten years up to a four-cylinder duplex Hoe printing-and-folding press, and saying to him: "Now, go ahead and find out how to run this machine, and print both sides of a signature upon it."

The average boy would shrink from the mechanical monster, and have no stomach whatever for "trial by error."

I think that the principle of determining the mind of a wild animal *along the lines of the professor* is not the best way. It should be developed *along the natural lines of the wild-animal mind.* It should be stimulated to do what it feels most inclined to do, and educated to achieve real mental progress.

I think that the ideal way to study the minds of apes, baboons and monkeys would be to choose a good location in a tropical or subtropical climate that is neither too wet nor too dry, enclose an area of five acres with an unclimbable fence, and divide it into as many corrals as there are species to be experimented upon. Each corral would need a shelter house and indoor playroom. The stage properties should be varied and abundant, and designed to stimulate curiosity as well as activity.

Somewhere in the program I would try to teach orang-utans and chimpanzees the properties of fire, and how to make and tend fires. I would try to teach them the seed-planting idea, and the meaning of seedtime and harvest. I would teach sanitation and cleanliness of habit, — a thing much more easily done than most persons suppose. I would teach my apes to wash dishes and to cook, and I am sure that some of them would do no worse than some human members of the profession who now receive $50 per month, or more, for spoiling food.

In one corral I would mix up a chimpanzee, an orang-utan, a golden baboon and a good-tempered rhesus monkey. My apes would begin at two years old, because after seven or eight years of age all apes are difficult, or even impossible, as subjects for peaceful experimentation.

I would try to teach a chimpanzee the difference between a noise and music, between heat and cold, between good food and bad food. Any trainer can teach an animal the difference between the blessings of peace and the horrors of war, or in other words, obedience and good temper versus cussedness and punishment.

Dr. Yerkes' laboratory in Montecito, California, and his experiments there with an orang-utan and other primates, were in a good place, and made a good beginning. It is very much to be hoped that means will be provided by which his work can be prosecuted indefinitely, and under the most perfect conditions that money can provide.

I hope that I will live long enough to see Dr. Yerkes develop the mind of a young grizzly bear in a four-acre lot, to the utmost limits of that keen and sagacious personality.

II

WILD ANIMAL TEMPERAMENT AND INDIVIDUALITY

In man and in vertebrate animals generally, temperament is the foundation of intelligence and progress. Fifty years ago Fowler and Wells, the founders of the science of phrenology and physiognomy, very wisely differentiated and defined four "temperaments" of mankind. The six types now recognized by me are the *morose, lymphatic, sanguine, nervous, hysterical* and *combative*; and their names adequately describe them.

This classification applies to the higher wild animals, quite as truly as to men. By the manager of wild animals in captivity, wild-animal temperament universally is recognized and treated as a factor of great practical importance. Mistakes in judging the temper of dangerous animals easily lead to tragedies and sudden death.

Fundamentally the temperament of a man or an animal is an inheritance from ancestors near or remote. In the human species a morose or hysterical temperament may possibly be corrected or improved, by education and effort. With animals this is rarely possible. The morose gorilla gives way to cheerfulness only when it is placed in ideally pleasant and stimulating social conditions. This, however, very seldom is possible. The nervous deer, bear or monkey is usually nervous to the end of its days.

The morose and hysterical temperaments operate against mental development, progress and happiness. In the human species among individuals of equal mental calibre, the sanguine individual is due to rise higher and go farther than his nervous or lymphatic rivals. A characteristic temperament may embrace the majority of a whole species, or be limited to a few individuals. Many species are permanently characterized by the temperament common to the majority of their individual members. Thus, among the great apes the gorilla species is either morose or lymphatic; and it is manifested by persistent inactivity and sullenness. This leads to loss of appetite, indigestion, inactivity and early death. Major Penny's "John Gorilla" was a notable exception, as will appear in Chapter IX.

The orang-utan is sanguine, optimistic and cheerful, a good boarder, affectionate toward his keepers, and friendly toward strangers. He eats well, enjoys life, lives long, and is well liked by everybody.

Except when quite young, the chimpanzee is either nervous or hysterical. After six years of age it is irritable and difficult to manage. After seven years of age (puberty) it is rough, domineering and dangerous. The male is given to shouting, yelling, shrieking and roaring, and when quite angry rages like a demon. I know of no wild animal that is more dangerous per pound than a male chimpanzee over eight years of age. When young they do wonders in trained performances, but when they reach maturity, grow big of arm and shoulder, and masterfully strong, they quickly become conscious of their strength. It is then that performing chimpanzees become unruly, fly into sudden fits of temper, their back hair bristles up, they stamp violently, and sometimes leap into a terrorized orchestra. Next in order, they are retired willy-nilly from the stage, and are offered for sale to zoological parks and gardens having facilities for confinement and control.

The baboons are characteristically fierce and aggressive, and in a wild state they live in troops, or even in herds of hundreds. Being armed with powerful canine teeth and wolf-like jaws, they are formidable antagonists, and other animals do not dare to attack them. It is because of their natural weapons, their readiness to fight like fiends, and their combined agility and strength that the baboons have been able to live on the ground and survive and flourish in lands literally reeking with lions, leopards, hyenas and wild dogs. The awful canine teeth of an old male baboon are quite as dangerous as those of any leopard, and even the leopard's onslaught is less to be feared than the wild rage of an adult baboon. In the Transvaal and Rhodesia, it is a common occurrence for an ambitious dog to go after a troop of baboons and never return.

Temperamentally the commoner groups of monkeys are thus characterized:

The rhesus monkeys of India are nervous, irritable and dangerous.

The green monkeys of Africa are sanguine, but savage and treacherous.

The langur monkeys of India are sanguine and peace-loving.

The macaques of the Far East vary from the sanguine temperament to the combative.

The gibbons vary from sanguine to combative.

The lemurs of Madagascar are sanguine, affectionate and peaceful.

Nearly all South American monkeys are sanguine, and peace-loving, and many are affectionate.

The species of the group of Carnivora are too numerous and too diversified to be treated with any approach to completeness. However, to illustrate this subject the leading species will be noticed.

TEMPERAMENTS OF THE LARGE CARNIVORES

The lion is sanguine, courageous, confident, reposeful and very reliable.

The tiger is nervous, suspicious, treacherous and uncertain.

The black and common leopards are nervous and combative, irreconcilable and dangerous.

The snow leopard is sanguine, optimistic and peace-loving. The puma is sanguine, good natured, quiet and peaceful.

The wolves are sanguine, crafty, dangerous and cruel.

The foxes are hysterical, timid and full of senseless fear.

The lynxes are sanguine, philosophic, and peaceful.

The mustelines are either nervous or hysterical, courageous, savage, and even murderous.

The bears are so very interesting that it is well worth while to consider the leading species separately. Possibly our conclusions will reveal some unsuspected conditions.

BEAR TEMPERAMENTS, BY SPECIES. The polar bears are sanguine, but in captivity they are courageous, treacherous and dangerous.

The Alaskan brown bears in captivity are sanguine, courageous, peaceful and reliable, but in the wilds they are aggressive and dangerous.

The grizzlies are nervous, keen, cautious, and seldom wantonly aggressive.

The European brown bears are sanguine, optimistic and good-natured.

The American black bears are sanguine and quiet, but very treacherous.

The sloth bears of India are nervous or hysterical, and uncertain.

The Malay sun bears are hysterical, aggressive and evil-tempered.

The Japanese black bears are nervous, cowardly and aggressive.

To those who form and maintain large collections of bears, involving much companionship in dens, it is necessary to keep a watchful eye on the temperament chart.

THE DEER. In our Zoological Park establishment there is no collection in which both the collective and the individual equation is more troublesome than the deer family. In their management, as with apes, monkeys and bears, it is necessary to take into account the temperament not only of the species, but also of each animal; and there are times when this necessity bears hard upon human nerves. The proneness of captive deer to maim and to kill themselves and each other calls for the utmost vigilance, and for heroic endurance on the part of the deer keeper.

Even when a deer species has a fairly good record for common sense, an individual may "go crazy" the instant a slightly new situation arises. We have seen barasingha deer penned up between shock-absorbing bales of hay seriously try to jump straight up through a roof skylight nine feet from the floor. We have seen park-bred axis deer break their own necks against wire fences, with 100 per cent of stupidity.

CHARACTERS OF DEER SPECIES

The white-tailed deer is sanguine, but in the fall the bucks are very aggressive and dangerous, and to be carefully avoided. The mule deer is sanguine, reasonable and not particularly dangerous.

The elk is steady of nerve, and sanguine in temperament, but in the rutting season the herd-masters are dangerous.

The fallow deer species has been toned down by a hundred generations of park life, and it is very quiet, save when it is to be captured and crated.

The axis deer is nervous, flighty, and difficult to handle.

The barasingha deer is hysterical and unaccountable.

The Indian and Malay sambar deer are lymphatic, confident, tractable and easily handled.

Never keep a deer as a "pet" any longer than is necessary to place it in a good home. All "pet deer" are dangerous, and should be confined all the time. Never go into the range or corral of a deer herd unless accompanied by the deer-keeper; and in the rutting season do not go in at all.

The only thoroughly safe deer is a dead one; for even does can do mischief. A SAMPLE OF NERVOUS TEMPERAMENT. As an example of temperament in small carnivores, we will cite the coati mundi of South America. It is one of the most nervous and restless animals we know. An individual of sanguine temperament rarely is seen. Out of about forty specimens with which we have been well acquainted, I do not recall one that was as quiet and phlegmatic as the raccoon, the nearest relative of *Nasua*. With a disposition so restless and enterprising, and with such vigor of body and mind, I count it strange that the genus *Nasua* has not spread all over our south-eastern states, where it is surely fitted to exist in a state of nature even more successfully than the raccoon or opossum.

The temper of the coati mundi is essentially quarrelsome and aggressive. While young, they are reasonably peaceful, but when they reach adult age, they become aggressive, and quarrels are frequent. Separations then are very necessary, and it is rare indeed that more than two adult individuals can be caged together. Even when two only are kept together, quarrels and shrill squealings are frequent. But they seldom hurt each other. The coati is not a treacherous animal, it is not given to lying in wait to make a covert attack from ambush, and being almost constantly on the move, it is a good show animal.

THE STRANGE COMBATIVE TEMPERAMENT OF THE GUANACO. In appearance the guanaco is the personification of gentleness. Its placid countenance indicates no guile, nor means of offense. Its lustrous gazelle-like eyes, and its soft, woolly fleece suggest softness of disposition. But in reality no animal is more deceptive. In a wild state amongst its own kind, or in captivity,—no matter how considerately treated,—it is a quarrelsome and at times intractable animal. "A pair of wild guanacos can often be seen or heard engaged in desperate combat, biting and tearing, and rolling over one another on the ground, uttering their gurgling, bubbling cries of rage. Of a pair so engaged, I shot one whose tail had then been bitten off in the encounter. In confinement, the guanaco charges one with his chest, or rears up on his hind legs to strike one with his fore-feet, besides biting and spitting up the contents of the stomach."—Richard Crawshay in "The Birds of Terra del Fuego."

MENTAL TRAITS AND TEMPER OF THE ATLANTIC WALRUS

Mr. Langdon Gibson, of Schenectady, kindly wrote out for me the following highly interesting observations on a remarkable arctic animal with which we are but slightly acquainted:

"In the summer of 1891, as a member of the first Peary Expedition I had an opportunity of observing some of the traits of the Atlantic walrus. I found him to be a real animal, of huge size, with an extremely disagreeable temper and most belligerently inclined. We hunted them in open whale-boats under the shadows of Greenland's mountain-bound coast, in the Whale Sound region, Lat. 77 degrees North.

"We hunted among animals never before molested, except by the Eskimo who (so far as I was able to ascertain) hunt them only during the winter season on the sea ice. We found animals whose courage and belief in themselves and their prowess had hitherto been unshaken by contact with the white man and his ingenious devices of slaughter.

"The walrus has a steady nerve and a thoroughly convincing roar. They have fought their kind and the elements for centuries and centuries, and know no fear. This, then, was the animal we sought in order to secure food for our dog teams. I can conceive of no form of big game hunting so conducive to great mental excitement and physical activity as walrus hunting from an open whale-boat. At the completion of such a hunt I have seen Eskimo so excited and worked up that they were taken violently sick with vomiting and headache.

"The walrus is a gregarious animal, confederating in herds numbering from ten to fifty, and in some instances no doubt larger numbers may be found together. On calm days they rest in unmolested peace on pans of broken ice which drift up and down the waters of Whale Sound. It is unfortunate that no soundings were taken in the region where the walrus were found, as a knowledge of the depth of water would have furnished some information as to the distances to which the animal will dive in search of food.

"The stomachs of all half- and full-grown walrus taken in Whale Sound were without exception well filled with freshly opened clams, with very few fragments of shells in evidence; the removal of the clam from the shell being as neatly accomplished as though done by an expert oysterman.

"In most cases these segregated herds of walrus were in charge of a large bull who generally occupied a central position in the mass of animals. Upon approaching such a herd for the first time, and when within about 200 feet, a large bull would lift his head, sniff audibly in our direction and give a loud grunt which apparently struck a responsive chord in the other sleeping animals. They would grunt in unison, in more subdued tones, after which the old walrus would drop his head to resume his interrupted nap. Their contempt for us was somewhat disconcerting.

"At the first crack of a rifle, however, the animals immediately aroused, and then during the fusillade which followed there occurred what might be called an orderly scramble for the water. In the first place the young ones were hustled to the edge of the ice-pan, and there, apparently under the protection of the mother's flipper, pushed into the water, immediately followed by the mother. The young bulls followed, and I recall no exceptions where the last animal into the water was not the big bull, who before diving would give our boat a wicked look and a roar of rage.

"The animals would immediately dive, and then we first became aware of a remarkable phenomenon. We found that when excited they would continue their roaring under water, and these strange sounds coming to us from below added considerably to the excitement of the chase. Although the cows and young animals would generally swim to places of safety, the other full grown animals would hover beneath our boat and from time to time come to the surface and charge. These charges were in all cases repulsed by the discharge of our rifles in the faces of the animals. The balls, however, from our .45 calibre carbines would flatten out under the skin on the massive bony structure of the animal's skull, and cause only a sort of rage and a sneeze, but it however had the effect of making them dive again. It is my belief that when enraged the walrus if not resisted would attack and attempt to destroy a boat. Icquah, one of our native hunters, showed me in the deck of his kyak two mended punctures which he told me were made by the tusks of a walrus that had made an *unprovoked* attack upon him.

"On more than one occasion I have seen two strong uninjured animals come to the assistance of a wounded companion, and swim away with it to a position of safety, *the injured animal being supported on both sides*, giving the appearance of three animals swimming abreast. The first time I witnessed this I did not comprehend its real meaning, but on another occasion in McCormick Bay I saw a wounded animal leaving a trail of blood and oil, supported on either side by two uninjured ones. They were making a hasty retreat and would occasionally dive together, but would quickly return to the surface.

"We found the most effective exposed spot to place a bullet was at the base of the animal's skull. A walrus instantly killed this way generally sinks, leaving a trail of blood and oil to mark the place of his descent. When hunting these animals it is well to have an Eskimo along with harpoon and line in readiness to make fast; otherwise one is apt to lose his quarry.

"In the early winter we usually found the walrus in smaller groups up in the bays. This was after the ice had begun to make, and in coming to the surface to breathe the animals found it necessary to butt their noses against the ice to break it. I have seen this done in ice at least four inches in thickness. In some instances I have seen a fractured star in the ice, a record of an unsuccessful attempt to make a breathing hole." Around these breathing holes we frequently found fragments of clam-shells, sections of crinoids and sea-anemones. It is evident that after raking the bottom with his tusks and filling his mouth with food, the walrus separates the food he desires to retain and rejects on his way up and at the surface such articles as he has picked up in haste and does not want.

"From the fact that the walrus is easily approached it is a simple matter to kill him with the modern high power rule. It is therefore to be hoped that future expeditions into the arctic seas will kill sparingly of these tremendous brutes which from point of size stand in the foremost rank among mammals."

The Elephant, Rhinoceros and Hippopotamus. *Individual Elephants* vary in temperament far more than do rhinoceroses or hippopotami, and the variations are wide. In a wild state, elephants are quiet and undemonstrative, almost to the point of dullness. They do not dominer, or hector, or quarrel, save when a rogue develops in the ranks, and sets out to make things interesting by the commission of lawless acts. A professional rogue is about everything that an orthodox elephant should not be, and he soon makes of himself so great a nuisance that he is driven out of the herd.

The temperament of the standardized and normal elephant is distinctly sanguine, *but a nervous or hysterical individual is easily developed by bad conditions or abuse.* Adult male elephants are subject to various degrees of what we may as well call sexual insanity, which is dangerous in direct proportion to its intensity. This causes many a "bad" show elephant to be presented to a zoological garden, where the dangers of this mental condition can at least be reduced to their lowest terms. Our Indian elephant who was known as Gunda was afflicted with sexual insanity, and he gradually grew worse, and increasingly dangerous to his keepers, until finally it was necessary to end his troubles painlessly with a bullet through his brain.

The Rhinoceros is a sanguine animal, of rather dull vision and slow understanding. In captivity it gives little trouble, and lives long. Adults individually often become pettish, or peevish, and threaten to prod their keepers without cause, but I have never known a keeper to take those lapses seriously. The average rhino is by no means a dull or a stupid animal, and they have quite enough life to make themselves interesting to visitors. In British East Africa a black rhinoceros often trots briskly toward a caravan, and seems to be charging, when in reality it is only desiring a "close-up" to satisfy its legitimate curiosity.

Every Hippopotamus, either Nile or pygmy, is an animal of serene mind and steady habits. Their appetites work with clock- like regularity, and require no winding. I can not recall that any one of our five hippos was ever sick for a day, or missed a meal. When the idiosyncrasies of Gunda, our bad elephant, were at their worst, the contemplation of Peter the Great ponderously and serenely chewing his hay was a rest to tired nerves. Keeper Thuman treats the four pygmy hippos like so many pet pigs,—save the solitary adult male, who sets himself up to be peevish. The breeding female is a wise and good mother, with much more maternal instinct than our chimpanzee "Suzette."

It may be set down as an absolute rule that hippos are lymphatic, easy-going, contented, and easy to take care of *provided* they are kept scrupulously clean, and are fed as they should be fed. They live long, breed persistently, give no trouble and have high exhibition value.

Giraffe individuals vary exceedingly,—beyond all other hoofed animals. Each one has its own headful of notions, and rarely will two be found quite alike in temperament and views of life. Some are sanguine and sensible, others are nervous, crotchety, and full of senseless fears. Those who are responsible for them in captivity are constantly harassed by fears that they will stampede in their stalls or yards, and break their own necks and legs in most unexpected ways. They require greater vigilance than any other hoofed animals we know. Sometimes a giraffe will develop foolishness to such a degree as to be unwilling to go out of its own huge door, into a shady and comfortable yard.

III

THE LANGUAGE OF WILD ANIMALS

Language is the means by which men and animals express their thoughts. Of language there are four kinds: vocal, pictured, written and sign language.

Any vocal sound uttered for the purpose of conveying thought, or influencing thought or action, is to be classed as vocal language. Among the mammals below man, *speech* is totally absent; but parrots, macaws, cockatoos and crows have been taught to imitate the sound of man's words, or certain simple kinds of music.

The primitive races of mankind first employed the sign language, and spoken words. After that comes picture language, and lastly the language of written words. Among the Indians and frontiersmen of the western United States and Canada, the sign language has reached what in all probability is its highest development, and its vocabulary is really wonderful.

The higher wild animals express their thoughts and feelings usually by sign language, and rarely by vocal sounds. Their power of expression varies species by species, or tribe by tribe, quite as it does among the races and tribes of men. It is our belief that there are today several living races of men whose vocabularies are limited to about 300 words.

Very many species of animals appear to be voiceless; but it is hazardous to attempt to specify the species. Sometimes under stress of new emergencies, or great pain, animals that have been considered voiceless suddenly give tongue. That hundreds of species of mammals and birds use their voices in promoting movements for their safety, there is no room to doubt. The only question is of the methods and the extent of voice used. Birds and men give expression to their pleasure or joy by singing.

In the jungle and the heavily wooded wilderness, one hears really little of vocal wild-animal language. Through countless generations the noisiest animals have been the first ones to be sought out and killed by their enemies, and only the more silent species have survived. All the higher animals, as we call the higher vertebrates, have the ability to exchange thoughts and convey ideas; and that is language.

At the threshold of this subject we are met by two interesting facts. Excepting the song-birds, the wild creatures of today have learned through instinct and accumulated experience that silence promotes peace and long life. The bull moose who bawls through a mile of forest, and the bull elk who bugles not wisely but too well, soon find their heads hanging in some sportsman's dining- room, while the silent Virginia deer, like the brook, goes on forever.

Association with man through countless generations has taught domestic animals not only the fact of their safety when giving voice, but also that very often there is great virtue in a vigorous outcry. With an insistent staccato neigh, the hungry horse jars the dull brain of its laggard master, and prompts him to "feed and water the stock." But how different is the cry of a lost horse, which calls for rescue. It cannot be imitated in printed words; but every plainsman knows the shrill and prolonged trumpet-call of distress that can be heard a mile or more, understandingly.

And think of the vocabulary of the domestic chicken! Years of life in fancied security have developed a highly useful vocabulary of language calls and cries. The most important, and the best known, are the following:

"Beware the hawk!" — "Coor! Coor!" "Murder! Help!" — "Kee- *owk*! Kee-*owk*! Kee-*owk*!" "Come on" — "Cluck! Cluck! Cluck!" "Food here! Food!" — "Cook-cook-cook-cook!" Announcement, or alarm — "Cut-cut-cut-*dah*-cut!" But does the wild jungle-fowl, the ancestor of our domestic chicken, indulge in all those noisy expressions of thought and feeling? By no means. I have lived for months in jungles where my hut was surrounded by jungle-fowl, and shot many of them for my table; but the only vocal sound I ever heard from their small throats was the absurdly shrill bantam-like crow of the cock. And even that led to several fatalities in the ranks of *Gallus stanleyi*.

Domestic cattle, swine and fowls have each a language of their own, and as far as they go they are almost as clear-cut and understandable as the talk of human beings. Just how much more is behind the veil that limits our understanding we cannot say; but no doubt there is a great deal.

But it is with the language of wild animals that we are most concerned. As already pointed out, wild creatures, other than songbirds, do not care to say much, because of the danger of attracting enemies that will exterminate them. Herein lies the extreme difficulty of ascertaining how wild beasts communicate. In the Animallai Hills of southern India I hunted constantly for many weeks through forests actually teeming with big game. There were herds upon herds of elephants, gaur, axis deer, sambar deer, monkeys by the hundred, and a good sprinkling of bears, wild hogs and tigers.

We saw hundreds upon hundreds of animals; but with the exception of the big black monkeys that used to swear at us, I can almost count upon my fingers the whole number of times that we heard animals raise their voices to communicate with each other.

Ape Voices. Naturally it is of interest to know something of the voices of the animals that physically and mentally stand nearest to man.

The wild gorilla has a voice almost equal to that of the chimpanzee, but in captivity he rarely utters any vocal sound other than a shriek, or scream.

The baby orang-utan either whines or shrieks like a human child. The half-grown or adult orang when profoundly excited bellows or roars, in a deep bass voice. Usually, however, it is a persistently silent animal.

The chimpanzee has a voice, and vociferously expresses its emotions.

First and most often is the plaintive, coaxing note, "Who'-oo! who'-oo! who'-oo!"

Then comes the angry and threatening, "Wah', wah', wah-! *Wah'*-hool *Wah'*-hool"

Lastly we hear the fearful, high-pitched yell or shriek, "Ah-h-h-h!" or "E-e-e-e."

The shriek, or scream, can be heard half a mile, and at close range it is literally ear-splitting. Usually it is accompanied by violent stamping or pounding with the feet upon the floor. It may signify rage, or nothing more than the joy of living, and of having a place in which to yell. It is this cry that is uncannily human-like in sound, and when heard for the first time it seems to register anguish.

In its Bornean jungle home, the orang-utan is nearly as silent as the grave. Never save once did I hear one utter a vocal sound. That was a deep bass roar emitted by an old male that I disturbed while he was sleeping on the comfortable nest of green branches that he had built for himself.

Concerning the chimpanzee, the late Mr. Richard L. Garner testified as follows:

"Not only does the chimpanzee often break the silence of the forest when all other voices are hushed, but he frequently answers the sounds of other animals, as if in mockery or defiance. ... Although diurnal in habit, the chimpanzees often make the night reverberate with the sounds of their terrific screaming, which I have known them to continue at times for more than an hour, with scarcely a moment's pause,—not one voice but many, and within the area of a square mile or so I have distinguished as many as seven alternating adult male voices.

"The gorilla is more silent and stoical than the chimpanzee, but he is far from being mute. He appears to be devoid of all risibility, but he is often very noisy. Although diurnal in habit, he talks less frequently during the day than at night, but his silence is a natural consequence of his stealth and cunning. There are times, however, when he ignores all danger of betraying his whereabouts or his movements, and gives vent to a deluge of speech. At night his screams and shouts are terrific."

The gibbons (including the siamang) have tremendous voices, with numerous variations, and they love to use them. My acquaintance with them began in Borneo, in the dense and dark coastal forest that there forms their home. I remember their cries as vividly as if I had heard them again this morning. While feeding, or quietly enjoying the morning sun, the gray gibbon (*Hylobates concolor*) emits in leisurely succession a low staccato, whistle-like cry, like "Hoot! Hoot! Hoot!" which one can easily counterfeit by whistling. This is varied by another whistle cry of three notes, thus: "Who-ee-hoo! Who-ee-hoo!" also to be duplicated by whistling. In hunting for specimens of that gibbon, for American museums, I could rarely locate a troop save by the tree-top talk of its members.

But all this was only childish prattle in comparison with the daily performances of the big white-handed, and the black hoolock gibbons, now and for several years past residing in our Primate House. Every morning, and perhaps a dozen times during the day, those three gibbons go on a vocal rampage and utter prolonged and ear-splitting cries and shrieks that make the welkin ring. The shrieking chorus is usually prolonged until it becomes tiresome to the monkeys. In all our ape and monkey experience we never have known its equal save in the vocal performances of Boma, our big adult male chimpanzee, the husband of Suzette.

A baboon emits occasionally, and without any warning, a fearful explosive bark, or roar, that to visitors is as startling as the report of a gun. The commonest expressions are "Wah!" and "*Wah'*-hoo!", and the visitor who can hear it close at hand without jumping has good nerves.

The big and solemn long-nosed monkey of Borneo (*Nasalis larvatus*) utters in his native tree-top (overhanging water), a cry like the resonant "honk" of a saxophone. He says plainly, "Kee honk," and all that I could make of its meaning was that it is used as the equivalent of "All's well."

Of all the monkeys that I have ever known, either wild or in captivity, the red howlers of the Orinoco, in Venezuela, have the most remarkable voices, and make the most remarkable use of them. The hyoid cartilage is expanded,—for Nature's own particular reasons,—into a wonderful sound-box, as big as an English walnut, which gives to the adult voice a depth of pitch and a booming resonance that is impossible to describe. The note produced is a prolonged bass roar, in alternately rising and falling cadence, and in reality comprising about three notes. It is the habit of troops of red howlers to indulge in nocturnal concerts, wherein four, five or six old males will pipe up and begin to howl in unison. The great volume of uncanny sound thus produced goes rolling through the still forest, far and wide; and to the white explorer who lies in his grass hammock in pitchy darkness, fighting off the mosquitoes and loneliness, and wondering from whence tomorrow's meals will come, the moral effect is gruesome and depressing.

In captivity the youthful howler habitually growls and grumbles in a way that is highly amusing, and the absurd pitch of the deep bass voice issuing from so small an animal is cause for wonder.

It is natural that we should look closely to the apes and monkeys for language, both by voice and sign. In 1891 there was a flood of talk on "the speech of monkeys," and it was not until about 1904 that the torrent stopped. At first the knowledge that monkeys can and do communicate to a limited extent by vocal sounds was hailed as a "discovery"; but unfortunately for science, nothing has been proved beyond the long-known fact that primates of a given species understand the meaning of the few sounds and cries to which their kind give utterance.

Thus far I have never succeeded in teaching a chimpanzee or orangutan to say even as much as "Oh" or "Ah." Nothing seems to be further from the mind of an orang than the idea of a new vocal utterance as a means to an end.

Our Polly was the most affectionate and demonstrative chimpanzee that I have ever seen, and her reaction to my voice was the best that I have found in our many apes. She knew me well, and when I greeted her in her own language, usually she answered me promptly and vociferously. Often when she had been busy with her physical- culture exercises and Delsartean movements on the horizontal bars or the trapeze in the centre of her big cage, I tested her by quietly joining the crowd of visitors in the centre of the room before her cage, and saying to her: "Polly! Wah! Wah! Wah!"

Nearly every time she would stop short, give instant attention and joyously respond "Wah! Wah! Wah!", repeating the cry a dozen times while she clambered down to the lower front bars to reach me with her hands. When particularly excited she would cry "*Who*-oo! *Who*-oo! *Who*-oo!" with great clearness and vehemence, the two syllables pitched four notes apart. This cry was uttered as a joyous greeting, and also at feeding-time, in expectation of food; but, simple as the task seems to be, I really do not know how to translate its meaning into English. In one case it appears to mean "How do you do?" and in the other it seems to stand for "Hurry up!"

Polly screamed when angry or grieved, just like a naughty child; and her face assumed the extreme of screaming-child expression. She whined plaintively when coaxing, or when only slightly grieved. With these four manifestations her vocal powers seemed to stop short. Many times I opened her mouth widely with my fingers, and tried to surprise her into saying "Ah," but with no result. It seems almost impossible to stamp the vocal-sound idea upon the mind of an orang-utan or chimpanzee. Polly uttered two distinct and clearly cut syllables, and it really seemed as if her vocal organs could have done more if called upon.

The cries of the monkeys, baboons and lemurs are practically nothing more than squeals, shrieks or roars. The baboons (several species, at least) bark or roar most explosively, using the syllable "Wah!" It is only by the most liberal interpretation of terms that such cries can be called language. The majority express only two emotions—dissatisfaction and expectation. Every primate calls for help in the same way that human beings do, by shrill screaming; but none of them ever cry "Oh" or "Ah."

The only members of the monkey tribe who ever spoke to me in their native forests were the big black langurs of the Animallai Hills in Southern India. They used to glare down at us, and curse us horribly whenever we met. Had we been big pythons instead of men they could not have said "Confound you!" any more plainly or more vehemently than they did.

In those museum-making days our motto was "All's fish that cometh to net"; and we killed monkeys for their skins and skeletons the same as other animals. My brown-skinned Mulcer hunters said that the bandarlog hated me because of my white skin. At all events, as we stalked silently through those forests, half a dozen times a day we would hear an awful explosion overhead, startling to men who were still-hunting big game, and from the middle zone of the tree-tops black and angry faces would peer down at us. They said: "Wah! Wah! Wah! Ah-^oo-oo-Aoo-oo-^oo-oo!" and it was nothing else than cursing and blackguarding. How those monkeys did hate us! I never have encountered elsewhere anything like it in monkeyland. la 1902 there was a startling exhibition of monkey language at our Primate House. That was before the completion of the Lion House. We had to find temporary outdoor quarters for the big jaguar, "Senor Lopez"; and there being nothing else available, we decided to place him, for a few days only, in the big circular cage at the north end of the range of outside cages. It was May, and the baboons, red-faced monkeys, rhesus, green and many other of the monkeys were in their outside quarters.

I was not present when Lopez was turned into the big: cage; but I heard it. Down through the woods to the polar bears' den, a good quarter of a mile, came a most awful uproar, made by many voices. The bulk of it was a medley of raucous yells and screeches, above which it was easy to distinguish the fierce, dog-like barks and roars of the baboons.

We knew at once that Lopez had arrived. Hurrying up to the Primate House, we found the wire fronts of the outside cages literally plastered with monkeys and baboons, all in the wildest excitement. The jaguar was in full view of them, and although not one out of the whole lot, except the sapajous, ever had an ancestor who had seen a jaguar, one and all recognized a hostile genus, and a hereditary enemy.

And how they cursed him, reviled him, and made hideous faces at him! The long-armed yellow baboons barked and roared until they were heard half a mile away. The ugly-tempered macaques and rhesus monkeys nearly burst with hatred and indignation. The row was kept up for a long time, and the monkey language that was lost to science on that occasion was, both in quantity and quality, beyond compare.

Bear Language. In their native haunts bears are as little given to loud talk as other animals; but in roomy and comfortable captivity, where many are yarded together, they rapidly develop vocal powers. Our bears are such cheerful citizens, and they do so many droll things, that the average visitor works overtime in watching them. I have learned the language of our bears sufficiently that whenever I hear one of them give tongue I know what he says. For example:

In warning or threatening an enemy, the sloth bear says: "Ach! Ach! Ach!" and the grizzly says: "Woof! Woof!" A fighting bear says: "Aw-aw-aw!" A baby's call for its mother is "Row! Row!" A bear's distress call is: "Err-*wow*-oo-oo-oof!"

But even in a zoological park it is not possible for everyone to recognize and interpret the different cries of bears, although the ability to do so is sometimes of value to the party of the second part. For example:

One day in February I was sitting in my old office in the Service Building, engrossed in I know not what important and solemn matter. The park was quiet; for the snow lay nine inches deep over all. There were no visitors, and the maintenance men were silently shovelling. Over the hill from the bear dens came the voice of a bear. It said, as plainly as print: "Err-wow!" I said to myself: "That sounds like a distress call," and listened to hear it repeated.

Again it came: "Err-wow!"

I caught up my hat and hastened over the hill toward the bear dens. On the broad concrete walk, about a hundred feet from the dens, four men were industriously shovelling snow, unaware that anything was wrong anywhere except on the pay-roll, opposite their names.

Guided by the cries that came from "The Nursery" den, where six yearling cubs were kept, I quickly caught sight of the trouble. One of our park-born brown bear cubs was hanging fast by one forefoot from the top of the barred partition. He had climbed to the top of the ironwork, thrust one front paw through between two of the bars (for bears are the greatest busybodies on earth), and when he sought to withdraw it, the sharp point of a bar in the overhang of the tree-guard had buried itself in the back of his paw, and held him fast. It seemed as if his leg was broken, and also dislocated at the shoulder. No wonder the poor little chap squalled for help. His mother, on the other side of the partition, was almost frantic with baffled sympathy, for she could do nothing to help him.

It did not take more than a quarter of a minute to have several men running for crowbars and other things, and within five minutes from the discovery we were in the den ready for action. The little chap gave two or three cries to let us know how badly it hurt his leg to hang there, then bent his small mind upon rendering us assistance.

First we lifted him up bodily, and held him, to remove the strain. Then, by good luck, we had at hand a stout iron bar with a U-shaped end; and with that under the injured wrist, and a crowbar to spring the treacherous overhang, we lifted the foot clear, and lowered little Brownie to the floor. From first to last he helped us all he could, and seemed to realize that it was clearly "no fair" to bite or scratch. Fortunately the leg was neither broken nor dislocated, and although Brownie limped for ten days, he soon was all right again.

After the incident had been closed, I gave the men a brief lecture on the language of bears, and the necessity of being able to recognize the distress call.

You can chase bison, elephants and deer all day without hearing a single vocal utterance. They know through long experience the value of silence.

The night after I shot my second elephant we noted an exception. The herd had been divided by our onslaught. Part of it had gone north, part of it south, and our camp for the night (beside the dead tusker) lay midway between the two. About bedtime the elephants began signalling to each other by trumpeting, and what they sounded was "The assembly." They called and answered repeatedly; and finally it became clear to my native followers that the two herds were advancing to unite, and were likely to meet in our vicinity. That particular trumpet call was different from any other I have ever heard. It was a regular "Hello" signal- call, entirely different from the "Tal-*loo*-e" blast which once came from a feeding herd and guided us to it.

But it is only on rare occasions that elephants communicate with each other by sound. I once knew a general alarm to be communicated throughout a large herd by the sign language, and a retreat organized and carried out in absolute silence. Their danger signals to each other must have been made with their trunks and their ears;

but we saw none of them, because all the animals were concealed from our view except when the two scouts of the herd were hunting for us.

In captivity an elephant trumpets in protest, or through fear, or through rage; but I am obliged to confess that as yet I cannot positively distinguish one from the other.

Once in the Zoological Park I heard our troublesome Indian elephant, Alice, roaring continuously as if in pain. It continued at such a rate that I hurried over to the Elephant House to investigate. And there I saw a droll spectacle. Keeper Richards had taken Alice out into her yard for exercise and had ordered her to follow him. And there he was disgustedly marching around the yard while Alice marched after him at an interval of ten paces, quite free and untrammeled, but all the while lustily trumpeting and roaring in indignant protest. The only point at which she was hurt was in her feelings.

Two questions that came into public notice concerning the voices of two important American animals have been permanently settled by "the barnyard naturalists" of New York.

The Voice of the American Bison. In 1907 the statement of George Catlin, to the effect that in the fall the bellowing of buffalo bulls on the plains resembled the muttering of distant thunder, was denied and severely criticized in a sportsman's magazine. On October 4 of that year, while we were selecting the fifteen bison to be presented to the Government, to found the Wichita National Bison Herd, four of us heard our best bull *bellow* five times, while another did the same thing four times.

The sound uttered was a deep-voiced roar,—not a grunt,—rising and falling in measured cadence, and prolonged about four or five seconds. It was totally different from the ordinary grunt of hunger, or the menace of an angry buffalo, which is short and sharp. In discussing the quality of the bellow, we agreed that it could properly be called a low roar. It is heard only in the rutting season,—the period described by Catlin,—and there is good reason to believe that Caitlin's description is perfectly correct.

The Scream of the Puma. This is a subject that will not lie still. I presume it will recur every five years as long as pumas endure. Uncountable pages of controversial letters have been expended upon the question: "Does the puma ever scream, like a woman in distress?"

The true answer is easy, and uncontestable by people whose minds are open to the rules of evidence.

Yes; the adult female puma DOES scream,-*in the mating season,* whenever it comes. It is loud, piercing, prolonged, and has the agonized voice qualities of a boy or a woman screaming from the pain of a surgical operation. To one who does not know the source or the cause, it is nerve-racking. When heard in a remote wilderness it must be truly fearsome. It says "Ow-w-w-w!" over and over. We have heard it a hundred times or more, and it easily carries a quarter of a mile.

The language of animals is a long and interesting subject,—so much so that here it is possible only to sketch out and suggest its foundations and scope. On birds alone, an entire volume should be written; but animal intelligence is a subject as far reaching as the winds of the earth.

No man who ever saw high in the heavens a V-shaped flock of wild geese, or heard the honk language either afloat, ashore or in the air, will deny the spoken language of that species. If any one should do so, let him listen to the wild-goose wonder tales of Jack Miner, and hear him imitate (to perfection) the honk call of the gander at his pond, calling to wild flocks in the sky and telling them about the corn and safety down where he is.

The woodpecker drums on the high and dry limb of a dead tree his resounding signal-call that is nothing more nor less (in our view) than so much sign language.

It was many years ago that we first heard in the welcome days of early spring the resounding *"Boo-hoo-hoo"* courting call of the cock pinnated grouse, rolling over the moist earth for a mile or more in words too plain to be misunderstood.

The American magpie talks beautifully; but I regret to say that I do not understand a word of its language. One summer we had several fine specimens in the great flying-cage, with the big and showy waterfowl, condor, griffon vulture, ravens and crows. One of those magpies often came over to the side of the cage to talk to me, and as I believe, make complaints. Whether he complained about his big and bulky cagemates, or the keepers, or me, I could not tell; but I thought that his grievances were against the large birds. Whenever I climbed over the guard rail and stooped down, he would come close up to the wire, stand in one spot, and in a quiet, confidential tone talk to me earnestly and gesticulate with his head for five minutes straight. I have heard senile old men run on in low-voiced, unintelligible clack in precisely the same way. The modulations of that bird's voice, its inflections and its vocabulary were wonderful. From his manner a messenger from Mars might easily have inferred that the bird believed that every word of the discourse was fully understood.

The lion roars, magnificently. The hyena "laughs"; the gray wolf gives a mournful howl, the coyote barks and howls, and the fox yaps. The elk bugles, the moose roars and bawls, in desire or defiance. The elephant trumpets or screams in the joy of good feeding, or in fear or rage; and it also rumbles deeply away down in its throat. The red squirrel barks and chatters, usually to scold some one whom he hates, but other small rodents know that silence is golden.

The birds have the best voices of all creatures. They are the sweet singers of the animal world, and to the inquiring mind that field is a wonderland.

The frogs are vociferous; and now if they were more silent they would last longer.

Of all the reptiles known to me, only two utter vocal sounds, — the alligator and the elephant tortoise. The former roars or bellows, the latter grunts.

IV

THE MOST INTELLIGENT ANIMALS

To the professional animal-man, year in and year out comes the eternal question, "Which are the most intelligent animals?"

The question is entirely legitimate. What animals are the best exponents of animal intelligence?

It seems to me that the numerous factors involved, and the comparisons that must be made, can best be expressed in figures. Opinions that are based upon only one or two sets of facts are not worth much. There are about ten factors to be taken into account and appraised separately.

In order to express many opinions in a small amount of space, we submit a table of estimates and summaries, covering a few mammalian species that are representative of many. But, try as they will, it is not likely that any two animal men will set down the same estimates. It all depends upon the wealth or the poverty of first-hand, eye-witness evidence. When we enter the field of evidence that must stand in quotation marks, we cease to know where we will come out. We desire to state that nearly all of the figures in the attached table of estimates are based upon the author's own observations, made during a period of more than forty years of ups and downs with wild animals. ESTIMATES OF THE COMPARATIVE INTELLIGENCE AND ABILITY OF CERTAIN CONSPICUOUS WILD ANIMALS, BASED UPON KNOWN PERFORMANCES, OR THE ABSENCE OF THEM. [Footnote: To the author, correspondence regarding the reasons for these estimates is impossible.]

[beginning of chart]

Perfection in all=100 [list of categories below are written vertically above the columns, with the last column unnamed and representing a total score of animal intelligence/1000]

Hereditary Knowledge Perceptive Faculties Original Thought Memory
Reason Receptivity in Training Efficiency in Execution Nervous

Energy Keenness of the Senses Use of the Voice

Primates

Chimpanzee 100 100 100 100 75 100 100 100 100 50 925
Orang-Utan 100 100 100 75 100 75 100 75 100 25 850
Gorilla. 50 50 50 50 75 25 25 50 100 25 500

Ungulates

Indian Elephant 100 100 100 100 100 100 100 75 50 25 850
Rhinoceros. 25 25 25 25 25 0 0 25 25 0 175
Giraffe 50 25 25 25 25 25 0 25 100 0 300
White-Tailed Deer . . .100 100 100 25 50 0 0 100 100 0 575
Big-Horn Sheep 100 100 50 25 50 0 0 100 100 0 525
Mountain Goat. 100 100 100 25 100 0 0 100 100 0 625
Domestic Horse. 100 100 100 75 75 75 75 100 100 50 850

Carnivores

Lion 100 100 50 75 50 75 50 100 100 25 725
Tiger 100 75 50 50 50 25 25 100 100 0 575
Grizzly Bear 100 100 50 25 50 75 50 75 100 25 725
Brown Bear (European)100 100 50 25 50 75 50 75 100 25 650
Gray Wolf 100 100 100 25 75 00 100 100 25 625
Coyote 100 75 50 25 50 0 0 75 100 25 500
Red Fox 100 100 50 75 100 0 0 100 100 25 650
Domestic Dog 50 100 75 75 75 75 100 100 100 100 850
Wolverine 100 100 100 25 100 0 75 100 100 0 700

Beaver 100 100 100 25 100 0 100 100 100 0 725

According to the author's information and belief, *these are "the most intelligent" animals:* The Chimpanzee is the most intelligent of all animals below man. His mind approaches most closely to that of man, and it carries him farthest upward toward the human level. He can learn more by training, and learn more easily, than any other animal.

The Orang-Utan is mentally next to the chimpanzee.

The Indian Elephant in mental capacity is third from man.

The high-class domestic Horse is a very wise and capable animal; but this is chiefly due to its age-long association with man, and education by him. Mentally the wild horse is a very different animal, and in the intellectual scale it ranks with the deer and antelopes.

The Beaver manifests, in domestic economy, more intelligence, mechanical skill and reasoning power than any other wild animal.

The Lion is endowed with keen perceptive faculties, reasoning ability and judgment of a high order, and its mind is surprisingly receptive.

The Grizzly Bear is believed to be the wisest of all bears.

The Pack Rat (*Neotona*) is the intellectual phenomenon of the great group of gnawing animals. It is in a class by itself.

The White Mountain Goat seems to be the wisest of all the mountain summit animals whose habits are known to zoologists and sportsmen.

A high-class Dog is the animal that mentally is in closest touch with the mind, the feelings and the impulses of man; and it is the only one that can read a man's feelings from his eyes and his facial expression.

The Marvelous Beaver. Let us consider this animal as an illuminating example of high-power intelligence.

In domestic economy the beaver is the most intelligent of all living mammals. His inherited knowledge, his original thought, his reasoning power and his engineering and mechanical skill in constructive works are marvelous and beyond compare. In his manifold industrial activities, there is no other mammal that is even a good second to him. He builds dams both great and small, to provide water in which to live, to store food and to escape from his enemies. He builds air-tight houses of sticks and mud, either as islands, or on the shore. When he cannot live as a pond-beaver with a house he cheerfully becomes a river-beaver. He lives in a riverbank burrow when house-building in a pond is impossible; and he will cheerfully tunnel under a stone wall from one-pond monotony, to go exploring outside.

[Illustration: CHRISTMAS AT THE PRIMATES' HOUSE Chimpanzees (with large ears) and orang-utans (small ears). The animal on the extreme right is an orang of the common caste]

He cuts down trees, both small and large, and he makes them fall as he wishes them to fall. He trims off all branches, and leaves no "slash" to cumber the ground. He buries green branches, in great quantity, in the mud at the bottom of his pond, so that in winter he can get at them under a foot of solid ice. He digs canals, of any length he pleases, to float logs and billets of wood from hinterland to pond.

If you are locating beavers in your own zoo, and are wise, you can induce beavers to build their dam where you wish it to be. This is how we did it!

We dug out a pond of mud in order that the beavers might have a pond of water; and we wished the beavers to build a dam forty feet long, at a point about thirty feet from the iron fence where the brook ran out. On thinking it over we concluded that we could manage it by showing the animals where we wished them to go to work.

We set a 12-inch plank on its edge, all the way across the dam site, and pegged it down. Above it the water soon formed a little pool and began to flow over the top edge in a very miniature waterfall. Then we turned loose four beavers and left them.

The next morning we found a cart-load of sticks and fresh mud placed like a dam against the iron fence. In beaver language this said to us:

"We would rather build our dam here, — if you don't mind. It will be easier for us, and quicker."

We removed all their material; and in our language that action said: "No; we would rather have you build over the plank."

The next night more mud and sticks piled against the fence said to us,

"We really *insist* upon building it here!"

We made a second clearance of their materials, saying in effect:

"You *shall not* build against the fence! You *must* build where we tell you!"

Thereupon, the beavers began to build over the plank, saying,

"Oh, well, if you are going to make a fuss about it, we will let you have your way."

So they built a beautiful water-tight dam precisely where we suggested it to them, and after that our only trouble was to keep them from overdoing the matter, and flooding the whole valley.

I am not going to dwell upon the mind and manners of the beaver. The animal is well known. Three excellent books have been written and pictured about him, in the language that the General Reader understands. They are as follows: "The American Beaver and His Works," Lewis H. Morgan (1868); "The Romance of the Beaver," A. R. Dugmore (no date); "History and Traditions of the Canada Beaver," H. T. Martin (1892).

"Clever Hans," the "Thinking Horse." From 1906 to 1910 the world read much about a wonderful educated horse owned and educated by Herr von Osten, in Germany. The German scientists who first came in touch with "Hans" were quite bowled over by the discovery that that one horse could "think." The *Review of Reviews* said, in 1910:

"It may be recalled that Clever Hans knew figures and letters, colors and tones, the calendar and the dial, that he could count and read, deal with decimals and fractions, spell out answers to questions with his right hoof, and recognize people from having seen their photographs. In every case his 'replies' were given in the form of scrapings with his right forehoof.

"Whether the questioner was von Osten, who had worked with him for seven years, or a man like Schillings, who was a complete stranger, seemed immaterial; and this went farthest, perhaps, in disposing of all talk of 'collusion' between master and beast."

Now, by the bald records of the case the fact was fixed for all time that Hans was the most wonderful mental prodigy that ever bore the form of a four-footed animal. His learning and his performances were astounding, and even uncanny. I do not care how he was trained, nor by what process he received ideas and reacted to them! He was a phenomenon, and I doubt whether this world ever sees his like again. His mastery of figures alone, no matter how it was wrought, was enough to make any animal or trainer illustrious.

But eventually Clever Hans came to grief. He was ostensibly thrown off his pedestal, in Germany, by human jealousy and egotism. Several industrious German scientists deliberately set to work to discredit him, and they stuck to it until they accomplished that task. The chief instrument in this was no less a man than the director of the "Psychological Institute" of the Berlin University, Professor Otto Pfungst. He found that when Hans was put on the witness stand and subjected to rigid cross examinations *by strangers*, his answers were due partly to *telepathy and hypnotic influence*! For example, the discovery was made that Hans could not always give the correct answer to a problem in figures unless it was known to the questioner himself.

To Hans's inquisitors this discovery imparted a terrible shock. It did not look like "thinking" after all! The mental process was *different* from the process of the German mind! The wonderful fact that Hans could remember and recognize and *reproduce* the ten digits was entirely lost to view. At once a shout went up all over Germany,—in the scientific circle, that Hans was an "impostor," that he could not "think," and that his mind was nothing much after all.

Poor Hans! The glory that should have been his, and imperishable, is gone. He was the victim of scientists of one idea, who had no sense of proportion. He truly WAS a thinking horse; and we are sure that there are millions of men whose minds could not be developed to the point that the mind of that "dumb" animal attained,—no, not even with the aid of hypnotism and telepathy.

The bare fact that a horse *can* be influenced by occult mental powers proves the close parallelism that exists between the brains of men and beasts. The Trap-Door Spider. Let no one suppose for one moment that animal mind and intelligence is limited to the brain-bearing vertebrates. The scope and activity of the notochord in some of the invertebrates present phenomena far more wonderful per capita than many a brain produces. Interesting books have been written, and more will be written hereafter, on the minds and doings of ants, bees, wasps, spiders and other insects.

Consider the ways and means of the ant-lion of the East, and the trap-door spider of the western desert regions. As one object lesson from the insect world, I will flash upon the screen, for a moment only, the trap-door spider. This wonderful insect personage has been exhaustively studied by Mr. Raymond L. Ditmars, in the development of a series of moving pictures, and at my request he has contributed the following graphic description of this spider's wonderful work.

"The trap-door spiders, inhabiting the warmer portions of both the Old and New Worlds, dig a deep tunnel in the soil, line this with a silken wallpaper, then construct a hinged door at the top so perfectly fitted and camouflaged with soil, that when it is closed there is no indication of the burrow. Moreover, the inside portion of the door of some species is so constructed that it may be "latched," there being two holes near the edge, precisely placed where the curved fangs may be inserted and the door held firmly closed. Also, the trap-door of a number of species is so designed as to be absolutely rain-proof, being bevelled and as accurately fitting a corresponding bevel of the tube as the setting of a compression valve of a gasolene engine.

[Illustration: THE TRAP-DOOR SPIDER'S DOOR AND BURROW By R. L. Ditmars 1. The door closed. Its top carefully counterfeits the surrounding ground. 2. The door with silken hinge, held open by a needle. 3. The spider in its doorway, looking for prey. 4. Section of the burrow and trap- door.]

"The study of a number of specimens of our southern California species, which builds the cork-type door, including observations of them at night, when they are particularly active, indicates that the construction of the tube involves other material than the silken lining employed by many burrowing spiders. In the excavation of the tube and retention of the walls, the spider appears to employ a glairy substance, which thoroughly saturates the soil and renders the interior of the tube of almost cement- like hardness. It is then plastered with a thick jet of silk from the spinning glands. This interior finishing process appears to be quite rapid, a burrow being readily lined within a couple of hours.

"The construction of the trap-door is a far more complicated process, this convex, beautifully bevelled entrance with its hinge requiring real scientific skill. Judging from observations on a number of specimens, the work is done from the outside, the spider first spinning a net-like covering over the mouth of the tube. This is thickened by weaving the body over the net, each motion leaving a smoky trail of silk. Earth is then shoveled into the covering, the spider carefully pushing the particles toward the centre, which soon sags, and assumes the proper curvature, and automatically moulds against the bevelled walls of the tube.

"The shoveling process must be nicely regulated to produce the proper bevel and thickness of the door. Then the cementing process is applied to the top, rendering the door a solid unit. From the actions of these spiders,—which often calmly rest an hour without a move,—it appears that the edges of the door are now subjected, by the stout and sharp fangs, to a cutting process like that of a can opener, leaving a portion of the marginal silk to act as a hinge. This hinge afterward receives some finishing touches, and the top of the door is either pebbled or finished with a few fragments of dead vegetation, cemented on, in order to exactly match the surrounding soil."

V

THE RIGHTS OF WILD ANIMALS

Every harmless wild bird and mammal has the right to live out its life according to its destiny; and man is in honor bound to respect those rights. At the same time it is a mistake to regard each wild bird or quadruped as a sacred thing, which under no circumstances may be utilized by man. We are not fanatical Hindus of the castes which religiously avoid the "taking of life" of any kind, and gently push aside the flea, the centipede and the scorpion. The reasoning powers of such people are strictly limited, the same as those of people who are opposed to the removal by death of the bandits and murderers of the human race.

The highest duty of a reasoning being is to reason. We have no moral or legal right to act like idiots, or to become a menace to society by protecting criminal animals or criminal men from adequate punishment. Like the tree that is known by its fruit, every alleged "reasoning being" is to be judged by the daily output of his thoughts.

Toward wild life, our highest duty is to be sane and sensible, in order to be just, and to promote the greatest good for the greatest number. Be neither a Hindu fanatic nor a cruel game- butcher like a certain wild-animal slaughterer whom I knew, who while he was on earth earned for himself a place in the hottest corner of the hereafter, and quickly passed on to occupy it.

The following planks constitute a good platform on which to base our relations with the wild animal world, and by which to regulate our duty to the creatures that have no means of defense against the persecutions of cruel men. They may be regarded as representing the standards that have been fixed by enlightened and humane civilization.

THE WILD ANIMALS' BILL OF RIGHTS

This Bill of Rights is to be copied and displayed conspicuously in all zoological parks and gardens, zoos and menageries; in all theatres and shows where animal performances are given, and in all places where wild animals and birds are trained, sold or kept for the pleasure of their owners.

Article 1. In view of the nearness of the approach of the higher animals to the human level, no just and humane man can deny that those wild animals have certain rights which man is in honor bound to respect.

Art. 2. The fact that God gave man "dominion over the beasts of the field" does not imply a denial of animal rights, any more than the supremacy of a human government conveys the right to oppress and maltreat its citizens.

Art. 3. Under certain conditions it is justifiable for man to kill a limited number of the so-called game animals, on the same basis of justification that domestic animals and fowls may be killed for food.

Art. 4. While the trapping of fur-bearing animals is a necessary evil, that evil must be minimized by reducing the sufferings of trapped animals to the lowest possible point, and by preventing wasteful trapping.

Art. 5. The killing of harmless mammals or birds solely for "sport," and without utilizing them when killed, is murder; and no good and humane man will permit himself to engage in any such offenses against good order and the rights of wild creatures.

Art. 6. Shooting at sea-going creatures from moving vessels, without any possibility of securing them if killed or wounded, is cruel, reprehensible, and criminal, and everywhere should be forbidden by ship captains, and also by law, under penalties.

Art. 7. The extermination of a harmless wild animal species is a crime; but the regulated destruction of wild pests that have been proven guilty, is sometimes necessary and justifiable.

Art. 8. No group or species of birds or mammals that is accused of offenses sufficiently grave to merit destruction shall be condemned undefended and unheard, nor without adequate evidence of a character which would be acceptable in a court of law.

Art. 9. The common assumption that every bird or mammal that offends, or injures the property of any man, is necessarily deserving of death, is absurd and intolerable. The death penalty should be the last resort, not the first one!

Art. 10. Any nation that fails adequately to protect its crop-and-tree-protecting birds deserves to have its fields and forests devastated by predatory insects.

Art. 11. No person has any moral right to keep a wild mammal, bird, reptile or fish in a state of uncomfortable, unhappy or miserable captivity, and all such practices should be prevented by law, under penalty. It is entirely feasible for a judge to designate a competent person as a referee to examine and decide upon each case.

Art. 12. A wild creature that cannot be kept in comfortable captivity should not be kept at all; and the evils to be guarded against are cruelly small quarters, too much darkness, too much light, uncleanliness, bad odors, and bad food. A fish in a glass globe, or a live bird in a cage the size of a collar-box is a case of cruelty.

Art. 13. Every captive animal that is suffering hopelessly from disease or the infirmities of old age has the right to be painlessly relieved of the burdens of life.

Art. 14. Every keeper or owner of a captive wild animal who through indolence, forgetfulness or cruelty permits a wild creature in his charge to perish of cold, heat, hunger or thirst because of his negligence, is guilty of a grave misdemeanor, and he should be punished as the evidence and the rights of captive animals demand.

Art. 15. An animal in captivity has a right to do all the damage to its surroundings that it can do, and it is not to be punished therefor.

Art. 16. The idea that all captive wild animals are necessarily "miserable" is erroneous, because some captive animals are better fed, better protected and are more happy in captivity than similar animals are in a wild state, beset by dangers and harassed by hunger and thirst. It is the opinion of the vast majority of civilized people that there is no higher use to which a wild bird or mammal can be devoted than to place it in perfectly comfortable captivity to be seen by millions of persons who desire to make its acquaintance.

Art. 17. About ninety-five per cent of all the wild mammals seen in captivity were either born in captivity or captured when in their infancy, and therefore have no ideas of freedom, or visions of their wild homes; consequently their supposed "pining for freedom" often is more imaginary than real.

Art. 18. A wild animal has no more inherent right to live a life of lazy and luxurious ease, and freedom from all care, than a man or woman has to live without work or family cares. In the large cities of the world there are many millions of toiling humans who are worse off per capita as to burdens and sorrows and joys than are the beasts and birds in a well kept zoological park. "Freedom" is comparative only, not absolute.

Art. 19. While the use of trained animals in stage performances is not necessarily cruel, and while training operations are based chiefly upon kindness and reward, it is necessary that vigilance should be exercised to insure that the cages and stage quarters of such animals shall be adequate in size, properly lighted and acceptably ventilated, and that cruel punishments shall not be inflicted upon the animals themselves.

Art. 20. The training of wild animals may, or may not, involve cruelties, according to the intelligence and the moral status of the trainer. This is equally true of the training of children, and the treatment of wives and husbands. A reasonable blow with a whip to a mean and refractory animal in captivity is not necessarily an act of cruelty. Every such act must be judged according to the evidence.

Art. 21. It is unjust to proclaim that "all wild animal performances are cruel" and therefore should be prohibited by law. The claim is untrue, and no lawmaker should pay heed to it. Wild animal performances are no more cruel or unjust than men-and-women performances of acrobatics. Practically all trained animals are well fed and tended, they welcome their performances, and go through them with lively interest. Such performances, when good, have a high educational value, — but not to closed minds.

Art. 22. Every bull-fight, being brutally unfair to the horses and the bull engaged and disgustingly cruel, is an unfit spectacle for humane and high-minded people, and no Christian man or woman can attend one without self-stultification.

Art. 23. The western practice of "bulldogging," now permitted in some Wild West shows, is disgusting, degrading, and never should be permitted.

Art. 24. The use of monkeys by organ-grinders is cruel, it is degrading to the monkeys, and should in all states be prohibited by law.

Art. 25. The keeping of live fishes in glass globes nearly always ends in cruelty and suffering, and should everywhere be prohibited by law. A round glass straight-jacket is just as painful as any other kind.

Art. 26. The sale and use of chained live chameleons as ornaments and playthings for idiotic or vicious men and children always means death by slow torture for the reptile, and should in all states be prohibited by law.

II. MENTAL TRAITS OF WILD ANIMALS

VI

THE BRIGHTEST MINDS AMONG AMERICAN ANIMALS

We repeat that *the most interesting features of a wild animal are its mind, its thoughts, and the results of its reasoning.* Besides these, its classification, distribution and anatomy are of secondary importance; but at the same time they help to form the foundation on which to build the psychology of species and individuals. Let no student make the mistake of concluding that when he has learned an animal's place in nature there is nothing more to pursue.

After fifty years of practical experience with wild animals of many species, I am reluctantly compelled to give the prize for greatest cunning and foresight *in self-preservation* to the common brown rat,—the accursed "domestic" rat that has adopted man as his perpetual servant, and regards man's goods as his lawful prey. When all other land animals have been exterminated from the earth, the brown rat will remain, to harry and to rob the Last Man.

The brown rat has persistently accompanied man all over the world. Millions have been spent in fighting him and the bubonic-plague flea that he cheerfully carries in his offensive fur. For him no place *that contains food* is too hot or too cold, too wet or too dry. Many old sailors claim to believe that rats will desert at the dock an outward-bound ship that is fated to be lost at sea; but that certificate of superhuman foreknowledge needs a backing of evidence before it can be accepted.

Of all wild animals, rats do the greatest number of "impossible" things. We have matched our wits against rat cunning until a madhouse yawned before us. Twice in my life all my traps and poisons have utterly failed, and left me faintly asking: *Are* rats possessed of occult powers? Once the answer to that was furnished by an old he-one who left his tail in my steel trap, but a little later *caught himself* in a trap-like space in the back of the family aeolian, and ignominiously died there, — a victim of his own error in judging distances without a tape line.

Tomes might be written about the minds and manners of the brown rat, setting forth in detail its wonderful intelligence in quickly getting wise to new food, new shelter, new traps and new poisons. Six dead rats are, as a rule, sufficient to put any *new* trap out of business; but poisons and infections go farther before being found out. [Footnote: For home use, my best rat weapon is rough-on-rats, generously mixed with butter and spread liberally on very thin slices of bread. It has served me well in effecting clearances.]

The championship for keen strategy in self-preservation belongs to the musk-oxen for their wolf-proof circle of heads and horns. Every musk-ox herd is a mutual benefit life insurance company. When a gaunt and hungry wolf-pack appears, the adult bull and cow musk-oxen at once form a close circle, with the calves and young stock in the centre. That deadly ring of lowered heads and sharp horns, all hung precisely right to puncture and deflate hostile wolves, is impregnable to fang and claw. The arctic wolves know this well. Mr. Stefansson says it is the settled habit of wolf packs of Banks Land to pass musk-ox herds without even provoking them to "fall in" for defense.

Judging by the facts that Charles L. Smith and the Norboe brothers related to Mr. Phillips and me around our camp-fires in the Canadian Rockies, the wolverine is one of the most cunning wild animals of all North America. This is a large order; for the gray wolf and grizzly bear are strong candidates for honors in that contest.

The greatest cunning of the wolverine is manifested in robbing traps, stealing the trapper's food and trap-baits, and at the same time avoiding the traps set for him. He is wonderfully expert in springing steel traps for the bait or prey there is in them, without getting caught himself. He will follow up a trap line for miles, springing all traps and devouring all baits as he goes. Sometimes in sheer wantonness he will throw a trap into a river, and again he will bury a trap in deep snow. Dead martens in traps are savagely torn from them. Those that can not be eaten on the spot are carried off and skilfully cached under two or three feet of snow.

Trapper Smith once set a trap for a wolverine, and planted close behind it a young moose skull with some flesh upon it. The wolverine came in the night, started at a point well away from the trap, dug a tunnel through six feet of snow, fetched up well behind the trap,—and triumphantly dragged away the head through his tunnel.

From the testimony of W. H. Wright, of Spokane, in his interesting book on "The Grizzly Bear," and for other reasons, I am convinced that the Rocky Mountain silver-tip grizzly is our brightest North American animal, and very keen of nose, eye, ear and brain. Mr. Wright says that "the grizzly bear far excels in cunning any other animal found throughout the Rocky Mountains, and, for that matter, he far excels them all combined." While the last clause is a large order, I will not dispute the opinion of a man of keen intelligence who has lived much among the most important and interesting wild animals of the Rockies.

In the Bitter Root Mountains Mr. Wright and his hunting party once set a bear trap for a grizzly, in a pen of logs, well baited with fresh meat. On the second day they found the pen demolished, the bait taken out, and everything that was movable piled on the top of the trap.

The trap was again set, this time loosely, under a bed of moss. The grizzly came and joyously ate all the meat that was scattered around the trap, but the moss and the trap were left untouched. And then followed a major operation in bear trapping. A mile away there was a steep slope of smooth rock, bounded at its foot by a creek. On one side was a huge tangle of down timber, on the other side loomed some impassable rocks; and a tiny meadow sloped away at the top. The half-fleshed carcasses of two dead elk were thrown half way down the rock slide, to serve as a bait. On the two sides two bear guns were set, and to their triggers were attached two long silk fish-lines, stretched taut and held parallel to each other, extending across the rocky slope. The idea was that the bear could not by any possibility reach the bait from above or below, without setting off at least one gun, and getting a bullet through his shoulders.

On the first night, no guns went off. The next morning it was found that the bear had crossed the stream and climbed straight up toward the bait until he reached the first fish-line; where he stopped. Without pressing the string sufficiently to set off its gun, he followed it to the barrier of trees. Being balked there, he turned about, retraced his steps carefully and followed the string to the barrier of rocks. Being blocked there, he back- tracked down the slide and across the stream, over the way he came. Then he widely circled the whole theatre, and came down toward the bait from the little meadow at its top of the slide.

Presently he reached the upper fish-line, twelve feet away from the first one. First he followed this out to the log barrier, then back to the rock ledge that was supposed to be unclimbable. There he scrambled up the "impossible" rocks, negotiated the ledge foot by foot, and successfully got around the end of line No. 2. Getting between the two lines he sailed out across the slope to the elk carcasses, feasted sumptuously, and then meandered out the way he came, without having disturbed a soul.

All this was done at night, and in darkness; and presumably that bear is there to this day, alive and well. No wonder Mr. Wright has a high opinion of the grizzly bear as a thinking animal.

In hiding their homes and young, either in burrows or in nests on the ground, wild rabbits and hares are wonderfully skilful, even under new conditions. Being quite unable to fight, or even to dig deeply, they are wholly dependent upon their wits in keeping their young alive by hiding them. Thanks to their keenness in concealment, the gray rabbit is plentiful throughout the eastern United States in spite of its millions of enemies. Is it not wonderful? The number killed by hunters last year in Pennsylvania was about 3,500,000!

The most amazing risk that I ever saw taken by a rabbit was made by a gray rabbit that nested in a shallow hole in the middle of a lawn-mower lawn east of the old National Museum building in Washington. The hollow was like that of a small wash-basin, and when at rest in it with her young ones the neutral gray back of the mother came just level with the top of the ground. At the last, when her young were almost large enough to get out and go under their own steam, a lawn-mower artist chanced to look down at the wrong moment and saw the family. Evidently that mother believed that the boldest ventures are those most likely to win.

Among the hoofed and horned animals of North America the white- tailed deer is the shrewdest in the recognition of its enemies, the wisest in the choice of cover, and in measures for self- preservation. It seems at first glance that the buck is more keen- witted than the doe; but this is a debatable question. Throughout the year the buck thinks only of himself. During fully one-half the year the doe is burdened by the cares of motherhood, and the paramount duty of saving her fawns from their numerous enemies. This, I am quite sure, is the handicap which makes it so much easier to kill a doe in the autumn hunting season than to bag a fully antlered and sophisticated buck who has only himself to consider.

The white-tailed deer saves its life by skulking low in timber and thick brush. This is why it so successfully resists the extermination that has almost swept the mule deer, antelope, white goat, moose and elk from all the hunting-grounds of the United States. Thanks to its alertness in seeing its enemies first, its skill and quickness in hiding, *and its mental keenness in recognizing and using deer sanctuaries,* the white- tailed or "Virginia" deer will outlive all the other hoofed animals of North America. In Pennsylvania they know enough to rush for the wire-bounded protected area whenever the hunters appear. That state has twenty-six such deer sanctuaries,— well filled with deer.

The moose and caribou dwell upon open or half-open ground, and are at the mercy of the merciless long-range rifles. Their keenness does not count much against rifles that can shoot and kill at a quarter of a mile. In the rutting season the bull moose of Maine or New Brunswick is easily deceived by the "call" of a birch-bark megaphone in the hands of a moose hunter who imitates the love call of the cow moose so skilfully that neither moose nor man can detect the falsity of the lure.

The mountain sheep is wide-eyed, alert and ready to run, but he dwells in exposed places from the high foothills up to the mountain summits, and now even the most bungling hunter can find him and kill him at long range. In the days of black powder and short ranges the sheep had a chance to escape; but now he has none whatever. He has keener vision and more alertness than the goat, but as a real life-saving factor that amounts to nothing! Wild sheep are easily and quickly exterminated.

The mountain goat has no protection except elevation and precipitous rocks, and to the hunter who has the energy to climb up to him he, too, is easy prey. Usually his biped enemy finds him and attacks him in precipitous mountains, where running and hiding are utterly impossible. When discovered on a ledge two feet wide leading across the face of a precipice, poor Billy has nothing to do but to take the bullets as they come until he reels and falls far down to the cruel slide-rock. He has a wonderful mind, but its qualities and its usefulness belong in Chapter XIII.

Warm-Coated Animals Avoid "Fresh Air." On this subject there is a strange divergence of reasoning power between the wild animals of cold countries and the sleeping-porch advocates of today.

Even the most warm-coated of the fur-bearing animals, such as the bears, foxes, beavers, martens and mink, and also the burrowing rodents, take great pains to den up in winter just as far from the "fresh air" of the cold outdoors as they can attain by deep denning or burrowing. The prairie-dog not only ensconces himself in a cul-de-sac at the end of a hole fourteen feet deep and long, but as winter sets in he also tightly plugs up the mouth of his den with moist earth. When sealed up in his winter den the black bear of the north draws his supply of fresh air through a hole about one inch in diameter, or less.

But the human devotees of fresh air reason in the opposite direction. It is now the regular thing for mothers to open wide to the freezing air of out-doors either one or all the windows of the rooms in which their children sleep, giving to each child enough fresh air to supply ten full-grown elephants, or twenty head of horses. And the final word is the "sleeping-porch!" It matters not how deadly damp is the air along with its 33 degrees of cold, or the velocity of the wind, the fresh air must be delivered. The example of the fat and heavily furred wild beast is ignored; and I just wonder how many people in the United States, old and young, have been killed, or permanently injured, by fresh air, during the last fifteen years.

And furthermore. Excepting the hoofed species, it is the universal rule of the wild animals of the cold-winter zones of the earth that the mother shall keep her helpless young close beside her in the home nest and keep them warm partly by the warmth of her own body. The wild fur-clad mother does not maroon her helpless offspring in an isolated cot in a room apart, upon a thin mattress and in an atmosphere so cold that it is utterly impossible for the poor little body and limbs to warm it and keep it warm. Yet many human mothers do just that, and some take good care to provide a warmer atmosphere for themselves than they joyously force upon their helpless infants.

No dangerous fads should be forced upon defenseless children or animals.

A proper amount of fresh air is very desirable, but the intake of a child is much less than that of an elephant. Besides, if Nature had intended that men should sleep outdoors in winter, with the moose and caribou, we would have been furnished with ruminant pelage and fat.

VII

KEEN BIRDS AND DULL MEN

If all men could know how greatly the human species varies from highest to lowest, and how the minds and emotions of the lowest men parallel and dove-tail with those of the highest quadrupeds and birds, we might be less obsessed with our own human ego, and more appreciative of the intelligence of animals.

A thousand times in my life my blood has been brought to the boiling point by seeing or reading of the cruel practices of ignorant and vicious men toward animals whom they despised because of their alleged standing "below man." By his vicious and cruel nature, many a man is totally unfitted to own, or even to associate with, dogs, horses and monkeys. Many persons are born into the belief that every man is necessarily a "lord of creation," and that all animals per se are man's lawful prey. In the vicious mind that impression increases with age. Minds of the better classes can readily learn by precept or by reasoning from cause to effect the duty of man to observe and defend the God- given rights of animals.

It was very recently that I saw on the street a group that represented man's attitude toward wild animals. It consisted of an unclean and vicious-looking man in tramp's clothing, grinding an offensive hand-organ and domineering over a poor little terrorized "ringtail" monkey. The wretched mite from the jungle was encased in a heavy woolen straight-jacket, and there was a strap around its loins to which a stout cord was attached, running to the Root of All Evil. The pavement was hot, but there with its bare and tender feet on the hot concrete, the sad-eyed little waif painfully moved about, peering far up into the faces of passers-by for sympathy, but all the time furtively and shrinkingly watching its tormentor. Every now and then the hairy old tramp would jerk the monkey's cord, each time giving the frail creature a violent bodily wrench from head to foot. I think that string was jerked about forty times every hour.

And that exhibition of monkey torture in a monkey hell continues in summer throughout many states of our country,—because "it pleases the children!" The use of monkeys with hand-organs is a cruel outrage upon the monkey tribe, and no civilized state or municipality should tolerate it. I call upon all humane persons to put an end to it.

As an antidote to our vaulting human egotism, we should think often upon the closeness of mental contact between the highest animals and the lowest men. In drawing a parallel between those two groups, there are no single factors more valuable than the home, and the family food supply. These hark back to the most primitive instincts of the vertebrates. They are the bedrock foundations upon which every species rests. As they are stable or unstable, good or bad, so lives or dies the individual, and the species also.

In employing the term "highest animals" I wish to be understood as referring to the warm-blooded vertebrates, and not merely the apes and monkeys that both structurally and mentally are nearest to man.

Throughout my lifetime I have been by turns amazed, entertained and instructed by the marvelous intelligence and mechanical skill of small mammals in constructing burrows, and of certain birds in the construction of their nests. Today the hanging nest of the Baltimore oriole is to me an even greater wonder than it was when I first saw one over sixty years ago. Even today the mechanical skill involved in its construction is beyond my comprehension. My dull brain can not figure out the processes by which the bird begins to weave its hanging purse at the tip end of the most unstable of all earthly building sites, — a down-hanging elm-tree branch that is swayed to and fro by every passing breeze. The situation is so "impossible" that thus far no moving picture artist has ever caught and recorded the process.

Take in your hand a standard oriole nest, and examine it thoroughly. First you will note that it is very strong, and thoroughly durable. It can stand the lashings of the fiercest gales that visit our storm-beaten shore.

How long would it take a man to unravel that nest, wisp by wisp, and resolve it into a loose pile of materials? Certainly not less than an entire day. Do you think that even your skilful fingers, — unassisted by needles, — could in two days, or in three, weave of those same materials a nest like that, that would function as did the original? I doubt it. The materials consist of long strips of the thin inner bark of trees, short strings, and tiny grass stems that are long, pliable and tough. Who taught the oriole how to find and to weave those rare and hard-to-find materials? And how did it manage all that weaving with its beak only? Let the wise ones answer, if they can; for I confess that I can not!

Down in Venezuela, in the delta of the Orinoco River, and elsewhere, lives a black and yellow bird called the giant cacique (pronounced cay-seek'), which as a nest-builder far surpasses our oriole. Often the cacique's hanging nest is from four to six feet long. The oriole builds to escape the red squirrels, but the cacique has to reckon with the prehensile-tailed monkeys.

Sometimes a dozen caciques will hang their nests in close proximity to a wasps' nest, as if for additional protection. A cacique's nest hangs like a grass rope, with a commodious purse at its lower end, entered by a narrow perpendicular slit a foot or so above the terminal facilities. It is impossible to achieve one of these nests without either shooting off the limb to which it hangs, or felling the tree. If it hangs low enough a charge of coarse shot usually will cut the limb, but if high, cutting it down with a rifle bullet is a more serious matter.

[Illustration with caption: HANGING NEST OF THE BALTIMORE ORIOLE
(From the "American Natural History")]

[Illustration with caption: GREAT HANGING NESTS OF THE CRESTED
CACIQUE As seen in the delta of the Orinoco Rover, Venezuela.]

To our Zoological Park visitors the African weaver birds are a wonder and a delight. Orioles and caciques do not build nests in captivity, but the weavers blithely transfer their activities to their spacious cage in our tropical-bird house. The bird-men keep them supplied with raffia grass, and they do the rest. Fortunately for us, they weave nests for fun, and work at it all the year round! Millions of visitors have watched them doing it. To facilitate their work the upper half of their cage is judiciously supplied with tree-branches of the proper size and architectural slant. The weaving covers many horizontal branches. Sometimes a group of nests will be tied together in a structure four feet long; and it branches up, or down, or across, seemingly without rhyme or reason.

Some of the weavers, which inhabit Africa, Malayana and Australia, are "communal" nest-builders. They build colonies of nests, close together. Imagine twenty-five or more Baltimore orioles massing their nests together on one side of a single tree, in a genuine village. That is the habit of some of the weaver birds;—and this brings us to what is called the most wonderful of all manifestations of house-building intelligence among birds. It is the community house of the little sociable weaver-bird of South Africa (*Philetoerus socius*). Having missed seeing the work of this species save in museums, I will quote from the Royal Natural History, written by the late Dr. Richard Lydekker, an excellent description: —This species congregates in large flocks, many pairs incubating their eggs under the same roof, which is composed of cartloads of grass piled on a branch of some camel- thorn tree in one enormous mass of an irregular umbrella shape, looking like a miniature haystack and almost solid, but with the under surface (which is nearly flat) honeycombed all over with little cavities, which serve not only as places for incubation, but also as a refuge against rain and wind.

"They are constantly being repaired by their active little inhabitants. It is curious that even the initiated eye is constantly being deceived by these dome-topped structures, since at a distance they closely resemble native huts. The nesting- chambers themselves are warmly lined with feathers."

Here must we abruptly end our exhibits of the intelligence of a few humble little birds as fairly representative of the wonderful mental ability and mechanical skill so common in the ranks of the birds of the world. It would be quite easy to write a volume on The Architectural Skill of Birds!

Now, let us look for a moment into the house-building intelligence and skill of some of the lower tribes of men. Out of the multitude of exhibits available I will limit myself to three, widely separated. In the first place, the habitations of the savage and barbaric tribes are usually the direct result of their own mental and moral deficiencies. The Eskimo is an exception, because his home and its location are dictated by the hard and fierce circumstances which dictate to him what he must do. Often he is compelled to move as his food supply moves. The Cliff-Dweller Indian of the arid regions of the Southwest was forced to cliff- dwell, in order to stave off extermination by his enemies. Under that spur he became a wonderful architect and engineer.

For present purposes we are concerned with three savage tribes which might have been rich and prosperous agriculturists or herdsmen had they developed sufficient intelligence to see the wisdom of regular industry.

Consider first the lowest of three primitive tribes that inhabit the extreme southern point of Patagonia, whose real estate holdings front on the Strait of Magellan. That region is treeless, rocky, windswept, cold and inhospitable. I can not imagine a place better fitted for an anarchist penal colony. North of it lie plains less rigorous, and by degrees less sterile, and finally there are lands quite habitable by cattle-and-crop-growing men.

But those three tribes elect to stick to the worst spot in South America. The most primitive is the tribe of "canoe Indians" of Tierra del Fuego, which probably represents the lowest rung of the human ladder. Beside them the cave men of 30,000 years ago were kings and princes. Their only rivals seem to be the Poonans of Central Borneo, who, living in a hot country, make no houses or shelters of any kind, and have no clothing but a long strip of bark cloth around the loins.

The Fuegians have long been known to mariners and travellers. They inhabit a region that half the year is bleak, cold and raw, but they make nothing save the rudest of the rude in canoes—of rough slabs tied together and caulked *with moss,*—and rough bone- pointed spears, bows, arrows and paddles. Their only clothing consists of skins of the guanacos loosely hung from the neck, and flapping over the naked and repulsive body. They make no houses, and on shore their only shelters from the wind and snow and chilling rains are rabbit-like forms of brush, broken off by hand.

These people are lower in the scale of intelligence than any wild animal species known to me; for they are mentally too dull and low to maintain themselves on a continuing basis. Their hundred years of contact with man has taught them little; and numerically they are decreasing so rapidly that the world will soon see the absolute finish of the tribe.

In the best of the three tribes, the Tchuelclus, the birth rate is so low that within recent times the tribe has diminished from about 5,000 to a remnant of about 500.

Now, have those primitive creatures "immortal souls?" Are they entitled to call chimpanzees, elephants, bears and dogs "lower animals?" Do they "think," or "reason," any more than the animals I have named?

It is a far cry from the highest to the lowest of the human race; and we hold that the highest animals intellectually are higher than the lowest men.

Now go with me for a moment to the lofty and dense tropical forest in the heart of the Territory of Selangor, in the Malay Peninsula. That forest is the home of the wild elephant, rhinoceros and sladang. And there dwells a jungle tribe called the Jackoons, some members of which I met at their family home, and observed literally in their own ancestral tree. Their house was not wholly bad, but it might have been 100 per cent better. It was merely a platform of small poles, placed like a glorified bird's nest in the spreading forks of a many-branched tree, about twenty feet from the ground. The main supports were bark-lashed to the large branches of the family tree. Over this there was a rude roof of long grass, which had a fairly intelligent slope. As a shelter from rain, the Jackoon house left much to be desired. The scanty loin cloths of the habitants knew no such thing as wash-day or line. With all its drawbacks, however, this habitation was far more adequate to the needs of its builders than the cold brush rabbit-forms of the Patagonian canoe Indians.

We now come to a tribe which has reduced the problem of housing and home life to its lowest common denominator. The Poonans of Central Borneo, discovered and described by Carl Bock, build *no houses of any kind,* not even huts of green branches; and their only overture toward the promotion of personal comfort in the home is a five-foot grass mat spread upon the sodden earth, to lie upon when at rest. And this, in a country where in the so-called "dry season" it rains half the time, and in the "wet season" all the time.

The Poonans have rudely-made spears for taking the wild pig, deer and smaller game, their clothes consist of bark cloth, around the loins only. They know no such thing as agriculture, and they live off the jungle.

It was said some years ago that a similarly primitive jungle tribe of Ceylon, known as the Veddahs, could count no more than five, that they could not comprehend "day after to-morrow," and that their vocabulary was limited to about 200 words.

It is very probable that the language of the Poonans and the Jackoons is equally limited. And what are we to conclude from all the foregoing? Briefly, I should say that the architectural skill of the orioles, the caciques and the weaver birds is greater than that of the South Patagonia native, the Jackoon and the Poonan. I should say that those bird homes yield to their makers more comfort and protection, and a better birth-rate, than are yielded by the homes of those ignorant, unambitious and retrogressive tribes of men now living and thinking, and supposed to be possessed of reasoning powers. If the whole truth could be known, I believe it would be found that the stock of ideas possessed and used by the groups of highly-endowed birds would fully equal the ideas of such tribes of simple-minded men as those mentioned. If caught young, those savages could be trained by civilized men, and taught to perform many tricks, but so can chimpanzees and elephants.

Curiously enough, it is a common thing for even the higher types of civilized men to make in home-building just as serious mistakes as are made by wild animals and savages. For example, among the men of our time it is a common mistake to build in the wrong place, to build entirely too large or too ugly, and to build a Colossal Burden instead of a real Home. From many a palace there stands forth the perpetual question: *"Why* did he do it?"

Any reader who at any time inclines toward an opinion that the author is unduly severe on the mentality of the human race, even as it exists today in the United States, is urged to read in the *Scientific Monthly* for January, 1922, an article by Professor L. M. Tennan entitled "Adventures in Stupidity. — A Partial Analysis of the Intellectual Inferiority of a College Student." He should particularly note the percentages on page 34 in the second paragraph under the subtitle "The Psychology of Stupidity."

VIII

THE MENTAL STATUS OF THE ORANG-UTAN

My first ownership of a live orang-utan began in 1878, in the middle of the Simujan River, Borneo, where for four Spanish dollars I became the proud possessor of a three-year old male. No sooner was the struggling animal deposited in the bottom of my own boat than it savagely seized the calf of my devoted leg and endeavored to bite therefrom a generous cross section. My leggings and my leech stockings saved my life. That implacable little beast never gave up; and two days later it died, — apparently to spite me.

My next orang was a complete reverse of No. 1. He liked not the Dyaks who brought him to me, but in the first moment of our acquaintance he adopted me as his foster-father, and loved me like a son. Throughout four months of jungle vicissitudes he stuck to me. He was a high-class orang, — and be it known that many orangs are thin-headed scrubs, who never amount to anything. His skull was wide, his face was broad, and he had a dome of thought like a statesman. He had a fine mind, and I am sure I could have taught him everything that any ape could learn.

During the four months that he lived with me I taught him, almost without effort, many things that were necessary in our daily life. Even the Dyaks recognized the fact that the "Old Man" was an orang (or "mias") of superior mind, and some of them traveled far to see him. Unfortunately the exigencies of travel and work compelled me to present him to an admiring friend in India. Mr. Andrew Carnegie and his then partner, Mr. J. W. Vandevorst, convoyed my Old Man and another small orang from Singapore to Colombo, Ceylon, whence they were shipped on to Madras, received there by my old friend A. G. R. Theobald, — and presented at the court of the Duke of Buckingham.

Up to a comparatively recent date, the studies of the psychologists that have been devoted to the minds of animals below man, have been chiefly concerned with low and common types. Comparatively few investigators have found it possible to make extensive and prolonged observations of the most intelligent wild animals of the world, even in zoological gardens, and their observations on wild animals in a state of nature seem to have been even more circumscribed. I know only three who have studied any of the great apes.

In attempting to fathom the mental capacity and the mental processes of some of the highest mammals, there is the same superior degree of interest attaching to the study of wild species that the ethnologist finds in the study of savage races of men that have been unspoiled by civilization. Obviously, it is more interesting to fathom the mind of a creature in an absolute state of nature than of one whose ancestors have been bred and reared in the trammels of domestication and for many successive generations have bowed to the will of man. The natural fury of the Atlantic walrus, when attacked, is much more interesting as a psychologic study than is the inbred rage of the bull-dog; and the remarkable defensive tactics of the musk-ox far surpass in interest the vagaries of range cattle.

For several reasons, the great apes, and particularly the chimpanzees and orang-utans, are the most interesting subjects for psychologic study of all the wild-animal species with which the writer is acquainted. Primarily this is due to the fact that intellectually and temperamentally, as well as anatomically, these animals stand very near to man himself, and closely resemble him. The great apes mentioned can give visible expression to a wide range of thoughts and emotions,

The voice of the adult orang-utan is almost absent, and only sufficient to display on rare occasions. What little there is of it, in animals over six years of age, is very deep and guttural, and may best be described as a deep-bass roar. Under excitement the orang can produce a roar by inhalation. Young orangs under two years of age often whine, or shriek or scream with anger, like excited human children, but with their larger growth that vocal power seems to leave them.

Despite the difference in temperament and quickness in delivery, I regard the measure of the orang-utan's mental capacity as being equal to that of the chimpanzee; but the latter is, and always will remain, the more alert and showy animal. The superior feet of the chimpanzee in bipedal work is for that species a great advantage, and the longer toes of the orang are a handicap. Although the orang's sanguine temperament is far more comforting to a trainer than the harum-scarum nervous vivacity of the chimpanzee, the value of the former is overbalanced, on the stage, by the superior acting of the chimp. For these reasons the trainers generally choose the chimp for stage education.

The chimpanzee is not only nervous and quick in thought and in action, but it is equally so *in temper.* It will play with any good friend to almost any extent, but the moment it suspects malicious unfairness, or what it regards as a "mean trick," it instantly becomes angry and resentful. Once when I attempted to take from our large black-faced chimpanzee, called Soko, a small lump of rubber which I feared she might swallow, my efforts were kindly but firmly thwarted. At last, when I diverted her by small offerings of chocolate, and at the right moment sought by a strategic movement to snatch the rubber from her, the palpable unfairness of the attempt caused the animal instantly to fly into a towering passion, and seek to wreak vengeance upon me. Her lips drew far back in a savage snarl, and she denounced my perfidy by piercing cries of rage and indignation. She also did her utmost to seize and drag me forcibly within reach of her teeth, for the punishment which she felt that I deserved.

A large male orang-utan named Dohong, under a similar test, revealed a very different mental attitude. He dexterously snatched a valuable watch-charm from a visitor who stood inside the railing of his cage, and fled with it to the top of his balcony. As quickly as possible I thrust my handkerchief between the bars, and waved it vigorously, to attract him. At once the animal came down to me, to secure another trophy, and before he realized his position I successfully snatched the charm from him, and restored it unharmed to its owner. Dohong seemed to regard the episode as a good joke. Without manifesting any resentment he turned a somersault on his straw, then climbed upon his trapeze and began to perform, as if nothing in particular had occurred.

The orang is distinctly an animal of more serene temper and more philosophic mind than the chimpanzee. This has led some authors erroneously to pronounce the orang an animal of morose and sluggish disposition, and mentally inferior to the chimpanzee. After a close personal acquaintance with about forty captive orangs of various sizes, I am convinced that the facts do not warrant that conclusion. The orang-utans of the New York Zoological Park certainly have been as cheerful in disposition, as fond of exercise and as fertile in droll performances as our chimpanzees. Even though the mind of the chimpanzee does act more quickly than that of its rival, and even though its movements are usually more rapid and more precise, the mind of the orang carries that animal precisely as far. Moreover, in its native jungles the orang habitually builds for itself a very comfortable nest on which to rest and sleep, which the chimpanzee ordinarily does not do.

I think that the exact mental status of an anthropoid ape is best revealed by an attempt to train it to do some particular thing, in a manner that the trainer elects. Usually about five lessons, carefully observed, will afford a good index of the pupil's mental capabilities. Some chimpanzees are too nervous to be taught, some are too obstinate, and others are too impatient of restraint. Some orang-utans are hopelessly indifferent to the business in hand, and refuse to become interested in it. I think that no orang is too dull to learn to sit at a table, and eat with the utensils that are usually considered sacred to man's use, but the majority of them care only for the food, and take no interest in the function. On the other hand, the average chimpanzee is as restless as a newly-caught eel, and its mind is dominated by a desire to climb far beyond the reach of restraining hands, and to do almost anything save that which is particularly desired.

Among the twenty or more orangs which up to 1922 have been exhibited in the Zoological Park, two stand out with special prominence, by reason of their unusual mental qualities. They differed widely from each other. One was a born actor and imitator, who loved human partnership in his daily affairs. The other was an original thinker and reasoner, with a genius for invention, and at all times impatient of training and restraint. The first was named Rajah, the latter was called Dohong.

Rajah was a male orang, and about four years of age when received by us. His high and broad forehead, large eyes and general breadth of cranium and jaw marked him at once as belonging to the higher caste of orangs. Dealers and experts have no difficulty in recognizing at one glance an orang that has a good brain and good general physique from those which are thin-headed, narrow-jawed, weak in body and unlikely to live long.

At the Zoological Park we have tested out the orang-utan's susceptibility to training, and proven that the task is so simple and easy that even amateurs can accomplish much in a short time. Desiring that several of our orangs should perform in public, we instructed the primate keepers to proceed along certain lines and educate them to that idea. Naturally, the performance was laid out to match our own possibilities. In a public park, where only a very little time can be devoted to training, we do not linger long over an animal that is either stupid or obstinate. Those which cannot be trained easily and quickly are promptly set aside as ineligible.

Without any great amount of labor, and with no real difficulty, our orangs were trained to perform the following simple acts:

1. To sit at table, and eat and drink like humans. This involved eating sliced bananas with a fork, pouring out milk from a teapot into a teacup, drinking out of a teacup, drinking out of a beer- bottle, using a toothpick, striking a match, lighting a cigarette, smoking and spitting like a man.

2. To ride a tricycle, or bicycle.

3. To put on a pair of trousers, adjust the suspenders, put on a sweater or coat, and a cap, reversing the whole operation after the performance.

4. To drive nails with a hammer.

5. Use a key to lock and unlock a padlock. The animal most proficient in this became able to select the right Yale key out of a bunch of half a dozen or more, with as much quickness and precision as the average man displays.

The orang Dohong learned to pedal and to guide a tricycle in about three lessons. He caught the two ideas almost instantly, and soon brought his muscles under control sufficiently to ride successfully, even under difficult conditions.

It was quickly recognized that our Rajah was a particularly good subject, and with him the keepers went farther than with the four others. From the first moment, the training operations were to him both interesting and agreeable. The animal enjoyed the work, and he entered into it so heartily that in two weeks he was ready to dine in public, somewhat after the manner of human beings.

A platform eight feet in height was erected in front of the Reptile House, and upon it were placed a table, a high chair such as small children use, and various dishes. To the platform a step- ladder led upward from the ground. Every day at four o'clock lusty Rajah was carried to the exhibition space, and set free upon the ground. Forthwith the keepers proceeded to dress him in trousers, vest, coat and cap. The moment the last button had been fastened and the cap placed upon his head, he would promptly walk to the ladder, climb up to the platform, and in the most business-like way imaginable, seat himself in his chair at the table, all ready to dine.

He used a napkin, ate his soup with a spoon, speared and conveyed his sliced bananas with his fork, poured milk from a teapot into his teacup, and drank from his cup with great enjoyment and decorum. When he took a drink (of tea) from a suspicious-looking black bottle, the audience always laughed. When he elevated the empty bottle to one eye and looked far into it, they roared; and when he finally took a toothpick and gravely placed it in his mouth, his auditors were delighted. Several times during the progress of each meal, Rajah would pause and benignly gaze down upon the crowd, like a self-satisfied judge on his bench.

Not once did Rajah spoil this exhibition, which was continued throughout an entire summer, nor commit any overt act of impatience, indifference or meanness. The flighty, nervous temper of the chimpanzee was delightfully absent. The most remarkable feature of it all was his very evident enjoyment of his part of the performance, and his sense of responsibility to us and to his audiences.

Rajah easily and quickly learned to ride a tricycle, and guide it himself. But for his untimely death, through a remarkable invasion of a microscopic parasite (*Balentidium coli*) imported from the Galapagos Islands by elephant tortoises, his mind would have been developed much farther. Since his death, in 1902, we have had other orang-utans that were successfully taught to dine, but none of them entered into the business with the same hearty zest which characterized Rajah, and made his performances so interesting.

We now come to a consideration of simian mental traits of very different character. Another male orang, named Dohong, of the same age and intellectual caste as Rajah, developed a faculty for mechanics and invention which not only challenged our admiration, but also created much work for our carpenters. He discovered, or invented, as you please, the lever as a mechanical force, — as fairly and squarely as Archimedes discovered the principle of the screw. Moreover, he delighted in the use of the new power thus acquired, quite as much as the successful inventor usually does. At the same time, two very bright chimpanzees of his own age, and with the same opportunities, discovered nothing.

[Illustration caption: THUMB-PRINT OF AN ORANG-UTAN
A group of fourteen experts in the New York City Departement of Criminal Records were unable to recognise this thumb print as anything else than that of a man]

[Illustration caption: "RAJAH," THE ACTOR ORANG-UTAN
In three lessons he learned to ride a tricycle]

Dohong was of a reflective turn of mind, and never was entirely willing to learn the things that his keepers sought to teach him. To him, dining at a table was tiresomely dull, and the donning of fashionable clothing was a frivolous pastime, On the other hand, the interior of his cage, and his gymnastic appliances of ropes, trapeze and horizontal bars, all interested him greatly. Every square inch of surface, and every piece of material in his apartment, was carefully investigated, many times over.

When three years old he discovered his own strength, and at first he used it good-naturedly to hector his cage-mate, a female chimpanzee smaller than himself. That, however, was of trifling interest. The day on which he made the discovery that he could break the wooden one and one-half inch horizontal bars that were held out from his cage walls on cast iron brackets, was for him a great day.

Before his discovery was noted by the keepers he had joyfully destroyed two bars, and with a broken piece used as a lever was attacking a third. These bars were promptly replaced by larger bars, of harder wood, but screwed to the same cast-iron brackets that had carried the first series.

For a time, the heavier bars endured; but in an evil moment the ape swung his trapeze bar, of two-inch oak, far over to one side of his cage, and applied the bar as a lever, inside of a horizontal bar and from above. The new force was too much for the cast-iron brackets, and one by one they gave way. Some were broken off, and others were torn from the wall by the breaking of the screws that held them. Knowing that all those brackets must be changed immediately, Dohong was left to destroy them; which he did, promptly and joyfully. We then made heavy brackets of flat wrought iron bars, 1/2 by 2 1/2 inches, unbreakable even with a lever. These were screwed on with screws so large and heavy that our carpenters knew they were quite secure.

[Illustration caption: THE LEVER THAT OUR ORANG-UTAN INVENTED, AND
THE WAY HE APPLIED IT By W. A. Camadeo, in the "Scientific American," 1907]

In due time, Dohong tested his lever upon the bars with their new brackets, and at first they held securely. Then he engaged Polly, his chimpanzee companion, to assist him to the limit of her strength. While Dohong pulled on the lever, Polly braced her absurd little back against the wall, and pushed upon it, with all her strength. At first nothing gave way. The combined strength exerted by the three brackets was not to be overcome by prying at the horizontal bar itself. It was then that Dohong's inventive genius rose to its climax. He decided to attack the brackets singly, and conquer them one by one. On examining the situation very critically, he found that each bracket consisted of a right- angled triangle of wrought iron, with its perpendicular side against the wall, its base uppermost, and its hypotenuse out in the air. Through the open centre of the triangle he introduced the end of his trapeze bar, chain and all, as far as it would go, then gave a mighty heave. The end of his lever was against the wall, and the power was applied in such a manner that few machine screws could stand so great a strain. One by one, the screws were torn out of the wood, and finally each bracket worked upon was torn off.

But there was one exception. The screws of one bracket were so firmly set in a particularly hard strip of the upright tongued- and-grooved yellow pine flooring that formed the wall, the board itself was finally torn out, full length! The board was four inches wide, seven-eighths of an inch thick, and seven feet long. Originally it was so firmly nailed that no one believed that it could be torn from its place. [Footnote: In the Winter of 1921 about a dozen newspapers in the United States published a sensational syndicated article, occupying an entire page, in which all of Dohong's lever discovery and cage-wrecking performances were reported as of recent occurrence, and credited to a stupid and uninteresting young orang called Gabong, now in the Zoological Park, that has not even the merit of sufficient intelligence to maintain a proper state of bodily uprightness, let alone the invention of mechanical principles.]

Without delay, Dohong started in with his lever to pry off the remaining boards of the wall, but this movement was promptly checked. Our next task consisted in making long bolts by which the brackets of the horizontal bars were bolted entirely through the partition walls and held so powerfully on the other side that even the lever could not wreck them.

As soon as the brackets were made secure, Dohong turned his attention to the two large sleeping boxes which were built very solidly on the balcony of his cage. Both of those structures he tore completely to pieces,—always working with the utmost good nature and cheerfulness. Realizing that they could not exist in the cage with him, we gave him a permit to tear them out—and save the time of the carpenters.

Dohong's use of his lever was seen by hundreds of visitors, and one frequent visitor to the Park, Mr. L. A. Camacho, an engineer, was so much impressed that he published in the *Scientific American* an illustrated account of what he saw.

For a long period, Dohong had been more or less annoyed by the fact that he could not get his head out between the front bars of his cage, and look around the partition into the home of his next-door neighbor. Very soon after he discovered the use of the lever, he swung his trapeze bar out to the upper corner of his cage, thrust the end of it out between the first bar and the steel column of the partition, and very deftly bent two of the iron bars outward far enough so that he could easily thrust his head outside and have his coveted look.

One of our later and largest orangs made a specialty of twisting the straw of his bedding into a rope six or seven feet long, then throwing it over his trapeze bar and swinging by it, forward and back.

Time and space will not permit the enumeration of the various things done by that ape of mechanical mind with his swinging rope and his trapeze, with ropes of straw *twisted by himself,* with keys, locks, hammer, nails and boxes. Any man who can witness such manifestations as those described above, and deny the existence in the animal of an ability to reason from cause to effect, must be prepared to deny the evidence of his own senses.

The individual variations between orangs, as also between chimpanzees, are great and striking. It may with truth be said that no two individuals of either species are really quite alike in physiognomy, temperament and mental capacity. As subjects for the experimental psychologist, it is difficult to see how any other could be found that would be even a good second in living interest to the great apes. The facts thus far recorded, so I believe, present only a suggestion of the rich results that await the patient scientific investigator. In the year 1915 Dr. Robert M. Yerkes, of Harvard University, conducted at Montecito, southern California, in a comfortable primate laboratory, six months of continuous and diligent experiments on the behavior of orang-utans and monkeys. His report, published under the title of "The Mental Life of Monkeys and Apes: A Study of Ideational Behavior," is a document of much interest and value. Dr. Yerkes' use of the orang-utan as a subject was a decided step forward in the study of "animal behavior" in America.

IX

THE MAN-LIKENESS OF THE CHIMPANZEE

During the past twenty years, millions of thinking people have been startled, and not a few shocked, by the amazing and uncanny human-likeness of the performances of trained chimpanzees on the theatrical stage. Really, when a well trained "chimp" is dressed from head to foot like a man, and is seen going with quickness, precision and spirit through a performance half an hour in length, we go away from it with an uncomfortable feeling that speech is all that he lacks of being a citizen.

In 1904 the American public saw Esau. Next came Consul,—in about three or four separate editions! In 1909 we had Peter. Then came I know not how many more, including the giant Casey and Mr. Garner's Susie; and finally in 1918 our own Suzette. The theatre-going public has been well supplied with trained chimpanzees, and the mental capacity of that species is now more widely known and appreciated than that of any other wild animal except the Indian elephant.

There are several reasons why chimpanzees predominate on the stage, and why so few performing orang-utans have been seen. They are as follows:

1. The orang is sanguine, and slower in execution than the nervous chimpanzee.

2. The feet of the orang are not good for shoes, and biped work.

3. The orang is rather awkward with its hands, and finally,

4. There are fully twice as many chimps in the market.

But the chimpanzee has certain drawbacks of his own. His nervous temper and his forced-draught activities soon wear him out. If he survives to see his sixth or seventh year, it is then that he becomes so strong and so full of ego that he becomes dangerous and requires to be retired.

Bright minds are more common among the chimpanzee species than among the orangs. Three chimps out of every five are good for training, but not more than two orangs out of five can be satisfactorily developed.

Some sensitive minds shrink from the idea that man has "descended" from the apes. I never for a moment shared that feeling. I would rather descend from a clean, capable and bright-minded genus of apes than from any unclean, ignorant and repulsive race of the genus *Homo*. In comparing the chimpanzees of Fernan Vaz with the Canoe Indians of the Strait of Magellan and other human tribes we could name, I think the former have decidedly the best of it. There are millions of members of the human race who are more loathsome and repulsive than wild apes.

The face of the chimpanzee is highly mobile, and the mouth, lips, eyes and voice express the various emotions of the individual with a degree of clearness and precision second only to the facial expression of man himself. In fact, the face of an intelligent chimpanzee or orang-utan is a fairly constant index of the state of mind of the individual. In their turn, those enormously expansive lips and keen brown eyes express contentment, doubt, fear and terror; affection, disapproval, jealousy, anger, rage; hunger and satiety; lonesomeness and illness.

The lips of the chimpanzee afford that animal several perfectly distinct expressions of the individual's mind and feelings. While it is not possible to offer a description of each which will certainly be recognizable to the reader, the two extremes will at least be appreciated. When coaxing for food, or attention, the lips are thrust far out beyond the teeth, and formed into a funnel with the small end outermost. When the chimpanzee flies into a rage at some real or fancied offense, the snarling lips are drawn back, and far up and down, until the teeth and gums are fully exposed in a ghastly threat of attack. At the same time, the voice gives forth shrill shrieks of rage, correctly represented by the syllable "Ee-ee-ee!", prolonged, and repeated with great force, three or four times. On such occasions as the latter, the offending party must look out for himself, or he may be roughly handled.

The voice of the chimpanzee is strong, clear, and in captivity it is very much in evidence. Two of its moderate tones are almost musical. It is heard when the animal says, coaxingly, "Who'-oe! Who'-oe!" A dozen times a day, our large specimens indulge in spells of loud yelling, purely for their own amusement. Their strident cry sounds like "Hoo-hoo-hoo-hoo! *Wah'*-hoo! *Wah'*-hoo! Hoo'-hoo! *Wah*-h-h-h! *Wah*-h-h!" The second combination, "Wah-hoo," consists of two sounds, four notes apart.

It is with their voices that chimpanzees first manifest their pleasure at seeing cherished friends of the human species, or their anger. Their recognition, and their exuberant joy on such occasions, is quite as apparent to every observer as are the manifestations of welcome of demonstrative human beings.

Like all other groups of species, the apes of various genera now living vary widely in their mentalities. The chimpanzee has the most alert and human-like mind but with less speed the orang-utan is a good second. The average captive gorilla, if judged by existing standards for ape mentality, is a poor third in the anthropoid scale, below the chimp and orang; but since the rise of Major Penny's family-pet gorilla, named John, we must revise all our former views of that species, and concede exceptions.

In studying the mental status of the primates I attach great importance to the work and results of the professional trainers who educate animals for stage performances. If the trainer does not know which are the brightest species of apes, baboons and monkeys, then who does? Their own fortunes depend upon their estimate of comparative mentality in the primates. Fortunately for our purposes, the minds of the most intelligent and capable apes, baboons, and monkeys have been partially developed and exploited by stage trainers, and to a far less extent by keepers in zoological parks. Some wonderful results have been achieved, and the best of these have been seen by the public in theatres, in traveling shows and in zoological parks. All these performances have greatly interested me, because they go so far as measures of mental capacity. I wish to make it clear that I take them very seriously.

[Illustration with caption: PORTRAIT OF A HIGH-CASTE CHIMPANZEE "Baldy" was an animal of fine intelligence and originality in thought. He was a natural comedian]

While many of the acts of trained animals are due to their power of mimicry and are produced by imitation rather than by original thought, even their imitative work reveals a breadth of intelligence, a range of memory and of activity and precision in thought and in energy which no logical mind can ignore. To say that a chimpanzee who can swing through thirty or forty different acts "does not think" and "does not reason," is to deny the evidence of the human senses, and fall outside the bounds of human reason.

Training Apes for Performances. As will appear in its own chapter, there is nothing at all mysterious in the training of apes. The subject must be young, and pliant in mind, and of cheerful and kind disposition. The poor subjects are left for cage life. The trainer must possess intelligence of good quality, infinite patience and tireless industry. Furthermore, the stage properties must be ample. An outfit of this kind can train any ape that is mentally and physically a good subject. Of course in every animal species, wild or domestic, there are individuals so dull and stupid that it is inexpedient to try to educate them.

The chimpanzee Suzette who came to us direct from the vaudeville stage performed every summer in her open-air "arena cage," until she entered motherhood, which put an end to her stage work. She was a brilliant "trick" bicycle rider. She could stand upright on a huge wooden ball, and by expert balancing and foot-work roll it up a steep incline, down a flight of stairs, and land it safely upon the stage, without once losing her balance or her control. She was entirely at home on roller skates, and when taken out upon the pavement of Baird Court she would go wildly careering around the large grass plat at high speed.

All the above acts were acrobatic feats that called for original thought and action, and were such as no dull mind and body could exert. All the training skill in the world could not take a machine and teach it to ride a bicycle through a collection of bottles, and an intelligent ape is a million years from being a "machine in fur and feathers."

More than once I have been astounded by the performances of apes on the stage. Mr. J. S. Edwards' orang-utan Joe was a very capable animal, and his performances were wonderful. He could use a hammer in driving nails, and a screwdriver in inserting and extracting screws, with wonderful dexterity.

The most remarkable chimpanzee performance that I ever saw was given in a New York theatre in 1909. The star actor was a fine male animal about six years old, called Peter. I made a complete record of his various acts, and the program was as follows

PERFORMANCE OF PETER, A CHIMPANZEE

Stage properties: a suit of clothes, shoes, chair, table, bed, bureau, hatrack, candle, cigarette, match, cuspidor, roller skates, bottles, flag, inclined plane and steps; plate, napkin, cup, spoon, teapot.

As Peter entered, he bowed to the audience, took off his cap and hung it upon a hatrack. He went to the table, seated himself in the chair, unfolded and put on a napkin, and with a string fastened it in place under his chin. With a fork he speared some slices of banana and ate them. Into his tumbler he poured liquid from a bottle, drank, then corked the bottle. Next, he poured tea into a cup, put in sugar and cream, took tea from the spoon, then drank from the cup. After that he took a toothpick and used it elaborately.

Striking a match he lit a cigarette, and smoked. In perfect man-fashion he took the cigarette between his fingers, gave his keeper a light, smoked again, and blew puffs of smoke first from one corner of his mouth and then the other. Then he elaborately spat into the cuspidor.

Next in order he went to the bureau, cleaned his teeth with a tooth-brush, brushed his hair on both sides, looked into the mirror and powdered his face.

Finally he bit a coin and put it on the keeper's plate as a tip.

He pulled off his coat, took off his cuffs and vest, and thus half undressed he joyously danced about, beating a tambourine. Then he removed his shirt, trousers, shoes, garters and socks. Lighting his candle he walked to his bed, blew out the candle and went to bed.

Very soon he rose, put on his trousers and a pair of roller skates and playfully pursued a young woman who ran before him. His use of the roller skates was excellent.

The stage was cleared of furniture, and a bicycle was brought out. He mounted it and started off, at the first trial, and swiftly rode around the stage about fifteen times. While riding he took off his cap and waved it. He rode up an inclined plane and down four steps without falling off, repeating for an encore, — but here he became peeved about something.

Five bottles were set in a figure 8, and he rode between them several times. At last he took up a bottle and drank out of it. Then he drank out of a tumbler, all while riding. After much flag- waving and swift riding, Peter stopped at the center of the stage, dismounted, bowed, clapped his hands vigorously and retired.

Peter's performance was remarkable because of the great length of it, the absolute skill and precision of it, and the animal's easy mastery of every situation. There was a notable absence of hesitations and mistakes, and of visible direction. The trainer seemed to do nothing save to assist with the stage properties, just as an assistant helps any acrobat through the property business of his act. If any commands or signs were given, the audience was not aware of it. Later on I learned that sometimes Peter did not perform with such spirit, and required some urging to be prompt. The trainer was kept hustling to keep up with his own duties. The animal seemed to remember, and I believe he did remember, the sequence of a performance of *fifty-six separate acts!*

When I witnessed Peter's performance in New York, saw the length of it and noted the immense amount of nervous energy that each performance used up, I made the prediction that he could not for one year endure such a strain. It was reported to me that he died nine months from that time.

In October, 1909, when Peter went to Philadelphia, he was frequently and closely studied and observed by Dr. Lightner Witmer, professor of psychology at the University of Pennsylvania, and his mentality was tested at the laboratory of the University. Dr. Witmer's conclusions, as set forth in a paper in the December (1909) issue of the *Psychological Clinic,* are of very great interest. He approached Peter's first performance in a skeptical frame of mind. I gladly waive the opportunity to express my own views regarding Peter in order to put upon the stand a more competent witness. Hear Dr. Witmer:

"As I entered the theatre," he says, "my feelings were commingled interest and doubt. My doubts were bred from knowledge of the difficulty of judging the intelligence of an animal from a stage performance. So-called educated horses and even educated seals and fleas have made their appeal in large number to the credulity of the public. Can any animal below man be educated in the proper sense of the word? Or is the animal mind susceptible of nothing more than a mechanical training, and only given the specious counterfeit of an educated intelligence when under the direct control of the trainer?

"Since that day I have seen Peter in five public performances, have tested him at my psychological clinic and privately on three occasions. I now believe that in a very real sense the animal is himself giving the stage performance. He knows what he is doing, he delights in it, he varies it from time to time, he understands the succession of tricks which are being called for, he is guided by word of mouth without any signal open or concealed, and the function of his trainer is exercised mainly to steady and control.

"I am prepared to accept the statement of his trainers, Mr. and Mrs. McArdle, that Peter's proficiency is not so much the result of training as of downright self-education."

Peter was put through many of the tests which Dr. Witmer uses for the study of backward children. He performed many of these tests in a very satisfactory manner. He was able to string beads the first time he tried it. He put pegs in the ordinary kindergarten pegging board. He opened and closed a very difficult lock. He used hammer and screw driver, and distinguished without any mistake between nails and screws. A peculiar kind of hammer was given to him in order to fool him, but Peter was not fooled. He felt both ends of the hammer and used the flat end instead of the round end.

Showing his initiative during the tests, Peter got away from those who were watching him and darted for a washstand, quickly turned the faucet and put his mouth to the spigot and secured a drink before he was snatched away by his trainers. He understood language and followed instructions without signs. He was able to say "mamma," and Doctor Witmer taught him in five minutes to give the sound of "p." The most remarkable performance was making the letter "w" on the blackboard, in which he imitated Doctor Witmer's movements exactly, and reproduced a fair copy of the letter.

The last four paragraphs reproduced above have been copied from an article which appeared in the Philadelphia *Public Ledger* on December 17, 1909.

Dr. Witmer declares that the study of this ape's mind is a subject fit, not for the animal psychologist, but for the child psychologist.

Suzette's Failure in Maternal Instinct. As a closing contribution to our observations on the chimpanzee, I must record a tragic failure in maternal instinct, as well as in general intelligence, in a chimpanzee.

In 1919 our two fine eight-year old chimpanzees, Boma and Suzette, were happily married. It was a genuine love match, and strictly monogamous at that; for while big Fanny Chimp in the cage next door to Boma loved Boma and openly courted him, he was outrageously indifferent to her, and even scorned her. After seven months of gestation, a very good baby was born to Suzette, quite naturally and successfully. Boma's shouts of excitement and delight carried half a mile throughout the Park. Everything looked most auspicious for the rearing of a wonderful cage-bred and cage-born chimpanzee, the second one ever born in captivity. Instead of carrying her infant astride her hip, as do orang mothers, and the coolie women of India, Suzette astonished us beyond measure by tucking it *into her groin,* between her thigh and her abdomen, head outward. It was a fine place, — warm and soft, — but not good when overdone! When Suzette walked, as she freely did, she held up the leg responsible for the baby, to hold it securely in place, and walked upon the other foot and her two hands. About all this there was one very bad thing. The baby was perfectly helpless! As long as the mother chose to keep it in her groin prison, it could not get free.

Suzette was completely isolated, kept absolutely quiet, and every chance was given her to go on with the functions of motherhood.

Her breasts contained plenty of milk, and the flow was due to start on the second day after the infant's arrival.

Day and night the baby was jealously confined in that massive and powerful groin,—and *under too much pressure!* When the baby cried, and kicked, and struggled to get free, Suzette would nervously rearrange her straw bed, carefully pick from the tiny fingers every straw that they had clutched, and settle down again. If the struggle was soon renewed, Suzette would change the infant over to the other groin, and close upon it as before.

Sleeping or waking, walking, sitting or lying down, she held it there. If we attempted to touch the infant, the mother instantly became savage and dangerous. Not one human finger was permitted to touch it. For hours, and for days, we anxiously watched for nursing to begin; but in vain. At last we became almost frantic from the spectacle of the infant being slowly starved to death because the mother did not realize that it needed her milk, and that she alone could promote nursing. *Her mother instinct utterly failed to supply the link that alone could connect infancy to motherhood, and furnish life.*

Of course this failure was due to poor Suzette's artificial life, and unnatural surroundings. Had she been all alone, in the depths of a tropical forest, Nature would have proceeded along her usual lines. But in our Primate House, Suzette felt that her infant was surrounded by a host of strange enemies, from whom it must be strongly and persistently *guarded and defended.* That was the idea that completely dominated her mind, ruled out all human help, and blocked the main process of nature.

During the eight days that the infant lived, it was able to reach her breast and nurse only once, for about one minute; and then back it went to its prison, where it died from sheer lack of nourishment.

In 1920, that same history was repeated, except that on this occasion our Veterinary Surgeon, Dr. W. Reid Blair, worked (on the fifth day) for seven hours without intermission to stupefy Suzette with chloroform, or other opiates, sufficiently to make it possible to remove the baby without a fight with the mother and its certain death. Owing to her savage temper all the work had to be done between iron bars, to keep from losing hands or arms, and the handicap on the human hand was too great. Even when Suzette had received chloroform for an hour and twenty minutes, and was regarded as *half dead,* at the first touch of a human finger upon her thigh she instantly aroused and sprang up, raging and ready for battle.

The whole effort failed. To rope Suzette and attempt to control her by force would have been sheer folly, or worse. In such a struggle the infant would have been torn to pieces.

The second one died as the first one did, and for an awful week we were unable to gain possession of the decomposing cadaver. Suzette knew that something was wrong, and she realized the awful odor, but that idea of defense of her offspring obscured all others. In maintaining her possession of that infant, nothing could surpass the cunning of that ape mother. Will we ever succeed in outwitting her, and in getting one of her babies alive into a baby incubator? Who can say?

X

THE TRUE MENTAL STATUS OF THE GORILLA

The true mental status of the gorilla was discovered in 1919 and 1920, at 15 Sloane Street, London, by Major Rupert Penny, of the Royal Air Service, and his young relative, Miss Alyse Cunningham. Prior to that time, through various combinations of retarding circumstances, no living gorilla had ever been placed and kept in an environment calculated to develop and display the real mental calibre of the gorilla mind. It seems that an exhibition cage, in a zoological park or garden thronged with visitors, actually tends to the suppression, or even the complete extinguishment, of true gorilla character. The atmosphere of the footlights and the stage in which the chimpanzee delights and thrives is to the gorilla repulsive and unbearable.

Judging by Major Penny's "John," the gorilla wishes to live in a high-class human family, in a modern house, and be treated like a human being! It is now definitely recognized by us, and also by our colleagues in the London Zoological Gardens, that gorillas can not live long and thrive on public exhibition, before great crowds of people, and that it is folly to insist upon trying to compel them to do so. The male individual that lived several years in the Breslau Zoological Garden and attained the age of seven years was a striking exception.

We have had two gorillas at our Park, one of which, a female named Dinah, arrived in good health, and lived with us eleven and one- half months. Her mind was dull and hopelessly unresponsive. She learned next to nothing, and she did nothing really interesting. Other captive gorillas I have known have been equally morose and unresponsive, and lived fewer months than Dinah.

It is because of such animals as Dinah that for fifty years the mental status of the gorilla species has been under a cloud. Until now it has been much misunderstood and unappreciated. Of the few gorillas that have been seen in England and America, I think that all save John have been so morose and unresponsive, *and so undeveloped by companionship and training*, that mentally they have been rated far below the chimpanzee and orang.

Our own Dinah was no exception to the rule. Personally she was a stupid little thing, even when in excellent health. Her most pronounced and exasperating stupidities were shown in her refusal to eat, or to taste, strange food, even when very hungry. Any ape that does not know enough to eat a fine, ripe banana, and will only mince away at the *inner lining* of the banana skin, is an unmitigated numskull, and hardly fit to live. Dinah was all that, and more. But, alas! We have seen a few stupid human children who obstinately refused even to taste certain new and unknown kinds of food, because they "know" they will not like them! So Dinah was not alone in her childish folly.

At last a chain of circumstances placed an intellectual and sensible gorilla, two years of age, in the hands of a family specially fitted by education and home surroundings to develop its mind and its manners. The results of those efforts have given to the gorilla an entirely new mental status. Thanks to the enterprise and diligence of Major Rupert Penny and Miss Cunningham in purchasing and caring for a sick and miserable young male gorilla,—a most hazardous risk,—a new chapter in wild-animal psychology now is to be written.

In December, 1918, "John Gorilla" was purchased in a London department store, out of a daily atmosphere heated to *85 degrees*, and a nightly condition of solitude and terror. From that awful state it was taken to live in Major Penny's comfortable apartments. John was seriously ill. He was in a "rickety" condition, and he weighed only 32 pounds. With a pure atmosphere, kept at 65 degrees only, and amid good surroundings, he soon became well. He attained such robust health and buoyant spirits that in March, 1921, he stood 40 1/2 inches high and weighed 112 pounds.

At my solicitation Miss Cunningham wrote out for me the very remarkable personal history of that wonderful animal,—apparently the most wonderful gorilla ever observed in captivity. It is a clear, straightforward and convincing record, and not one of its statements is to be for one moment doubted. While it is too long to reproduce here in its entirety, I will present a condensation of it, in Miss Cunningham's own words that will record the salient facts,— with no changes save in arrangement.

Miss Cunningham says:

LONELINESS. "We soon found it was impossible to leave him alone at night, because he shrieked every night, and nearly all night, from loneliness and fear. This we found he had done in the store where he lived before coming to us. He always began to cry directly he saw the assistants putting things away for the night. We found that this loneliness at night was trying on his health and appetite. As soon as possible my nephew had his bed made up every night in the room adjoining the cage, with the result that John was quite happy, and began to grow and put on fat.

TREATMENT. "I fed him, washed his hands, face and feet twice a day, and brushed and combed his hair,—which he would try to do himself whenever he got hold of the brush or comb. He soon got to like all this.

TRAINING. "My next idea was to teach him to be strictly clean in his habits. It was my ambition to be able to have him upstairs in our house as an ordinary member of the household. I taught him first as a child is taught and handled. This took some time. At first I could not make him understand what we expected of him, even though I always petted him and gave him grapes (of which he was especially fond), but I think at first he imagined that this treatment was a punishment. At first, without other reasons, he would roll on the floor and shriek, but directly he understood what was expected of him he soon learned, and began to behave excellently.

"This training occupied quite six weeks. About February, 1919, we took him out of his cage, and allowed him the freedom of the house. Thereafter he would run upstairs to the bathroom of his own accord, turning the doorknob of whatever room he was in, and also opening the door of the bathroom.... He would get out of bed in the night by himself, go back to bed, and pull the blankets over himself quite neatly.

FOOD. "John's appetite seemed to tire of foods very quickly. The only thing he stuck to was milk, which he liked best when warmed. We began by giving him a quart a day, rising to three and one-half quarts a day. I found that he preferred to choose his own food, so I used to prepare for him several kinds, such as bananas, oranges, apples, grapes, raisins, currants, dates and any small fruits in season, such as raspberries or strawberries, *all of which he liked to have warmed!*

"These displays I placed on a high shelf in the kitchen, where he could get them with difficulty. I think that he thought himself very clever when he stole anything. He never would eat anything stale. He was extremely fond of fresh lemon jelly, but he never would touch it after the second day. He loved roses, *to eat*, more than anything. The more beautiful they were, the more he liked them, but he never would eat faded roses. He never cared much for nuts of any other kind than baked peanuts, save walnuts. I found that nuts gave him dreadful spells of indigestion.

USE OF TOOLS. "He knew what hammers and chisels were for, but for obvious reasons we never encouraged him in anything to do with carpentry. With cocoanuts he was very funny. He knew that they had to be broken, and he would try to break them on the floor. When he found he couldn't manage that, he would bring the nut to one of us and try to make us understand what he wished. If we gave him a hammer he would try to use it on the nut, and on not being able to manage that, he would give back to us both the hammer and the cocoanut.

GAMES AND PLAY. "We never taught him any tricks; he simply acquired knowledge himself. A game he was very fond of was to pretend he was blind, shutting his eyes very tightly, and running around the room knocking against tables and chairs.... We found that exercise was the thing he required to keep him in health, and my nephew used to give him plenty of that by playing hide and seek with him in the morning before breakfast, and in the evening before dinner, — up and down stairs, in and out of all the rooms. He simply loved that game, and would giggle and laugh while being chased…. If he saw that a stranger was at all nervous about him, he loved running past him, and giving him a smack on the leg, — and you could see him grin as he did so.

"A thing he greatly enjoyed was to stand on the top rail of his bed and jump on the springs, head over heels, just like a child.

CAUTION. "He was very cautious. He would never run into a dark room without first turning on the light.

FEAR. "John seemed to realize danger for other people in high places, for if anyone looked out of a high window he always pushed them away if he were at the window himself, but if he was away from it he would run and pull them back.... He was very much afraid of full-grown sheep, cows and horses, but he loved colts, calves and lambs, proving to us that he recognized youth.

WOODS VS. FIELDS. "We found he did not like fields or open country, but he was very happy in a garden, or in woods.... He always liked nibbling twigs, and to eat the green buds of trees.

TABLE MANNERS. "His table manners were really very good. He always sat at the table, and whenever a meal was ready, would pull his own chair up to his place. He did not care to eat a great deal, but he especially liked to drink water out of a tumbler.... He was the least greedy of all the animals I have ever seen. He never would snatch anything, and always ate very slowly. He always drank a lot of water, which he would always get himself whenever he wanted it by turning on a tap. Strange to say, he always turned off the water when he had finished drinking.

PLAYING TO THE GALLERY. "John seemed to think that everyone was delighted to see him, and he would throw up the window whenever he was permitted. If he found the sash locked he would unfasten it, and when a big crowd had collected outside he would clap his chest and his hands. [Footnote: In the summer of 1920 a globe-trotter just arrived from England excitedly reported to me: "While driving along a street in London *I saw a live gorilla* in the upper window of an apartment. It was a *real gorilla;* and it clapped its hands at us as we looked! Now *what* did it all mean?" Fortunately I was able to explain it.]

PUNISHMENT AND REPENTANCE. "We made one very great mistake with John. His cage was used as a punishment, with the result that we never could leave him there alone, for he would shriek all the time.... Now, a stick was the one thing that our gorilla would not stand from anyone, save Major Penny and myself. Presently we found out that the only way to deal with him was to tell him that he was very naughty, and push him away from us; when he would roll on the floor and cry, and be very-repentant, holding one's ankles, and putting his head on our feet.

AFFECTION FOR A CHILD. "He was especially fond of my little niece, three years old. John and she used to play together for hours, and he seemed to understand what she wanted him to do. If she ever cried, and her mother would not go and pick her up, John would always try and nip the mother, or give her a smack with the full weight of his hand, evidently thinking she was the cause of the child's tears.

A SENSE OF GOOD ORDER. "He loved to take everything out of a wastepaper basket and strew the contents all over the room, after which, when told to do so he would pick up everything and put it all back, but looking very bored all the while. If the basket was very full he would push it all down very carefully, to make room for more. He would always put things back when told to do so, such as books from a bookshelf or things from a table.

[Illustration caption: THE GORILLA WITH THE WONDERFUL MIND Owned by Major Rupert Penny, educated by Miss Alyse Cunningham, London, 1918-1921]

TWO CASES OF ORIGINAL THOUGHT. (1) "One day we were going out, for which I was sitting ready dressed, when John wished to sit in my lap. My sister, Mrs. Penny, said: 'Don't let him. He will spoil your dress.'

"As my dress happened to be a light one I pushed him away, and said, 'No!' He at once lay on the floor and cried just like a child, for about a minute. Then he rose, looked round the room, found a newspaper, went and picked it up, spread it on my lap and climbed up. This was quite the cleverest thing I ever saw him do. *Even those who saw it said they would not have believed it had they not seen it themselves!* Both my nephews, (Major Penny and Mr. E. C. Penny), his wife and my sister (Mrs. Penny) were in the room, and can testify to the correctness of the above record.

(2) "Another clever thing John did, although I suspect this was due more to instinct that to downright cleverness. A piece of filet beefsteak had just come from the butcher. Inasmuch as occasionally I gave him a small mouthful of raw beef, a small piece of the coarser part of the steak was cut off, and I gave it to him. He tasted it, then gravely handed it back to me. Then he took my hand and put it on the finer part of the meat. From that I cut off a tiny piece, gave it to him, and he ate it. When my nephew came home he wouldn't believe it, so I tried it again, with the same result, except that then he did not even attempt to eat the coarser meat."

* * * * *

Concerning Miss Cunningham's wonderful story, I wish to state that I believe all of it, — because there is no reason to do otherwise! It sets a new mark in gorilla lore, and it lifts a curtain from an animal mind that previously was unknown, and very generally misunderstood.

To the Doubting Thomases who will doubt some portions of Miss Cunningham's story, let me cite, by way of caution, the following history:

When Du Chaillu discovered the gorilla, and came to America and England with his specimens to tell about it, he said that when a big gorilla is attacked and made angry it beats its breast, repeatedly, with its clenched fists. The wiseacres of that day solemnly shook their heads and said: "Oh, no! That can not be true. No ape ever did that. He is romancing!" But now we know that this breast-beating and chest-clapping habit is to a gorilla a common-place performance, even in captivity.

Sometimes there are more things in heaven and earth than are dreamt of in all our philosophy.

XI

THE MIND OF THE ELEPHANT

It was in the jungles of the Animallai Hills of southern India that I first became impressed by the mental capacity of the Indian elephant. I saw many wild herds. I saw elephants at work, and at one period I lived in a timber camp, consisting of working elephants and mahouts. I saw a shrewd young elephant-driver soundly flogged for stealing an elephant, farming it out to a native timber contractor for four days, and then elaborately pretending that the animal had been "lost." Later on I saw elephant performances in the "Greatest Show on Earth" and elsewhere, and for eighteen years I have been chief mourner over the idiosyncrasies of Gunda and Alice. If I do not now know something about elephants, then my own case of animal intelligence is indeed hopeless.

To me it seems that the only thing necessary to establish the elephant as an animal of remarkable intellect and power of original reasoning is to set forth the unadorned facts that lie ready to hand.

Cuvier recorded the opinion that in sagacity the elephant in no way excels the dog and some other species of carnivora. Sir Emerson Tennent, even after some study of the elephant, was disposed to award the palm for intelligence to the dog, but only "from the higher degree of development consequent on his more intimate domestication and association with man." In the mind of G. P. Sanderson we fear that familiarity with the elephant bred a measure of contempt; and this seems very strange. He says:

"Its reasoning faculties are undoubtedly far below those of the dog, and possibly of other animals; and in matters beyond its daily experience it evinces no special discernment."

To me it seems that all three of those opinions are off the target. The dog is not a wild, untrammeled animal; and neither dogs, cats nor savage men evince any special discernment "beyond the range of their daily experience." Moreover, there are some millions of tame men of whom the same may be said with entire safety.

Very often the question is asked: "Is the African elephant equal in intelligence and training capacity to the Indian species?"

To this we must answer: Not proven. We do not know. The African species never has been tried out on the same long and wide basis as the Indian. Many individual African elephants, very intelligent, have been trained, successfully, and have given good accounts of themselves. For my own part I am absolutely sure that when taken in hand at the same age, and trained on the same basis as the Indian species, the African elephant will be found mentally quite the equal of the Indian, and just as available for work or performances.

No negro tribe really likes to handle elephants and train them. The Indian native loves elephants, and enjoys training them and working with them. It is these two conditions that have left the African elephant far behind the procession. The African elephant belongs to the great Undeveloped Continent. He has been, and he still is, mercilessly pursued and slaughtered for his tusks. All the existing species of African elephants are going down and out before the ivory hunters. We fear that they will all be dead one hundred years from this time, or even less. A century hence, when the last *africanus* has gone to join the mammoth and the mastodon, his well protected wild congener in India still will be devouring his four hundred pounds of green fodder per day, and the tame ones will be performing to amuse the swarming human millions of this overcrowded world.

In the minds of our elephant keepers, familiarity with elephants has bred just the reverse of contempt. Both Thuman and Richards are quite sure that elephants are the wisest of all wild animals.

Despite the very great amount of trouble made for Keeper Thuman by Gunda, the Indian, and Kartoum, the African, Thuman grows enthusiastic over the shrewdness of their "cussedness." He is particularly impressed by their skill in opening chain shackles, and unfastening the catches and locks of doors and gates. And really, Kartoum's ingenuity in finding out how to open latches and bolts is almost inexhaustible, as well as marvelous.

Keeper Richards declares that our late African pygmy elephant, Congo, was the wisest animal he ever has known. I have elsewhere referred to his ability in shutting his outside door. Richards taught him to accept coins from visitors, deposit them in a box, then pull a cord to ring a bell, one pull for each coin represented. The keeper devised four different systems of intimate signals by which he could tell Congo to stop at the right point, and all these were so slight that no one ever detected them. One was by a voice-given cue, another by a hand motion, and a third was by an inclination of the body.

Keeper Richards relates that Congo would go out in his yard, collect a trunkful of peanuts from visitors, bring them inside and secretly cache them in a corner behind his feed box. Then he would go out for more graft peanuts, bring them in, hide them and proceed to eat the first lot. There are millions of men who do not know what it is to conserve something that can be eaten.

In this discussion of the intellectual powers and moral qualities of the elephant I will confine myself to my own observations on *Elephas indicus*, except where otherwise stated. A point to which we ask special attention is that in endeavoring to estimate the mental capacity of the elephant, we will base no general conclusions upon *any particularly intelligent individual*, as all mankind is tempted to do in discussions of the intelligence of the dog, the cat, the horse, parrot and ape. On the contrary, it is our desire to reveal the mental capacity of *every elephant living*, tame or wild, except the few individuals with abnormal or diseased minds. It is not to be shown how successfully *an* elephant has been taught by man, but how *all* elephants in captivity have been taught, and the mental capacity of *every* elephant.

Under the head of intellectual qualities we have first to consider the elephant's

POWERS OF INDEPENDENT OBSERVATIONS, AND REASONING FROM CAUSE TO EFFECT

While many wonderful stories are related of the elephant's sagacity and independent powers of reasoning, it must be admitted that a greater number of more wonderful anecdotes are told on equally good authority of dogs. But the circumstances in the case are wholly to the advantage of the universal dog, and against the rarely seen elephant. While the former roams at will through his master's premises, through town and country, mingling freely with all kinds of men and domestic animals, with unlimited time to lay plans and execute them, the elephant in captivity is chained to a stake, with no liberty of action whatever aside from begging with his trunk, eating and drinking. His only amusement is in swaying his body, swinging one foot, switching his tail, and (in a zoological park) looking for something that he can open or destroy. Such a ponderous beast cannot be allowed to roam at large among human beings, and the working elephant never leaves his stake and chain except under the guidance of his mahout. There is no means of estimating the wonderful powers of reasoning that captive elephants might develop if they could only enjoy the freedom accorded to all dogs except the blood-hound, bull-dog and a few others.

In the jungles of India the writer frequently has seen wild elephants reconnoitre dangerous ground by means of a scout or spy; communicate intelligence by signs; retreat in orderly silence from a lurking danger, and systematically march, in single file, like the jungle tribes of men.

Once having approached to within fifty yards of the stragglers of a herd of about thirty wild elephants, which was scattered over about four acres of very open forest and quietly feeding, two individuals of the herd on the side nearest us suddenly suspected danger. One of them elevated his trunk, with the tip bent forward, and smelled the air from various points of the compass. A moment later an old elephant left the herd and started straight for our ambush, scenting the air with upraised trunk as he slowly and noiselessly advanced. We instantly retreated, unobserved and unheard. The elephant advanced until he reached the identical spot where we had a moment before been concealed. He paused, and stood motionless as a statue for about two minutes, then wheeled about and quickly but noiselessly rejoined the herd. In less than half a minute the whole herd was in motion, heading directly away from us, and moving very rapidly, but *without the slightest noise*. The huge animals simply vanished like shadows into the leafy depths of the forest. Before proceeding a quarter of a mile, the entire herd formed in single file and continued strictly in that order for several miles. Like the human dwellers in the jungle, the elephants know that the easiest and most expeditious way for a large body of animals to traverse a tangled forest is for the leader to pick the way, while all the others follow in his footsteps.

In strong contrast with the stealthy and noiseless manner in which elephants steal away from a lurking danger, or an ambush discovered, from an open attack accompanied with the noise of firearms they rush away at headlong speed, quite regardless of the noise they make. On one occasion a herd which I was designing to attack, and had approached to within forty yards, as its members were feeding in some thick bushes, discovered my presence and retreated so silently that they had been gone five minutes before I discovered what their sudden quietude really meant. In this instance, as in several others, the still alarm was communicated by silent signals, or sign-language.

At the Zoological Park we reared an African pygmy elephant (*Elephas pumilio*). When his slender little tusks grew to eighteen inches in length he made some interesting uses of them. Once when the keepers wished to lead him upon our large platform scales, the trembling of the platform frightened him. He conceived the idea that it was unsafe, and therefore that he must keep off. He backed away, halted, and refused to leave solid ground. The men pushed him. He backed, and trumpeted a shrill protest. The men pushed harder, and forced him forward. Trumpeting his wild alarm and his protest against what he regarded as murder, he fell upon his knees and drove his tusks into the earth, quite up to his mouth, to anchor himself firmly to the solid ground. It was pathetic, but also amusing. When Congo finally was pushed upon the scales and weighed, he left the trembling instrument of torture with an air of disgust and disapproval that was quite as eloquent as words. On several occasions when taken out for exercise in the park, he endeavored to hinder the return to quarters by anchoring himself to Mother Earth.

Congo once startled us by his knowledge of the usefulness of doors. For a time he was kept in a compartment that had an outside door running sidewise on a trolley track, and controlled by two hanging chains, one to close it and one to open it. Each chain had on its end a stout iron ring for a handle. One chilly morning when I went to see Congo, I asked his keeper to open his door, so that he could go out.

The keeper did so, by pulling the right hand chain. The moment the draft of chilly outer air struck Congo, who stood in the centre of his stall facing me, he impatiently wheeled about, walked up to the left hand chain, grabbed it with his trunk, slipped the ring over one of his tusks, then inclined his head downward and with an irritated tug pulled the door shut with a spiteful slam. "Open it again," I said to the keeper.

He did so, and in the same way, but with a visible increase in irritation, Congo closed it in the same manner as before. Again the keeper opened the door, and this time, with a real exhibition of temper Congo again thrust the ring over his tusk, and brought the door shut with a resounding bang. It was his regular habit to close that door, or to open it, when he felt like more air or less air; and who is there who will say that the act was due to "instinct" in a jungle-bred animal, or anything else than original thought. The ring on his tusk was his own invention, as a means to a desired end.

Every elephant that we ever have had has become, through his own initiative and experimenting, an expert in unfastening the latches of doors and gates, and in untying chains and ropes. Gunda always knew enough to attack the padlocks on his leg chains, and break them if possible. No ordinary clevis would hold him. When the pin was threaded at one end and screwed into its place, Gunda would work at it, hour by hour, until he would start it to unscrewing, and then his trunk-tip would do the rest. The only clevis that he could not open was one in which a stout cotter pin was passed through the end of the clevis-pin and strongly bent.

Through reasons emanating in his own savage brain, Gunda took strong dislikes to several of our park people. He hated Dick Richards,—the keeper of Alice. He hated a certain messenger boy, a certain laborer, a painter and Mr. Ditmars. Toward me he was tolerant, and never rushed at me to kill me, as he always did to his pet aversions. He stood in open fear of his own keeper, Walter Thuman, until he had studied out a plan to catch him off his guard and "get him." Then he launched his long-contemplated attack, and Thuman was almost killed.

Our present (1921) male African elephant, Kartoum, is not so hostile toward people, but his insatiable desire is to break and to smash all of his environment that can be bent or broken. His ingenuity in finding ways to damage doors and gates, and to bend or to break steel beams, is amazing. His greatest feat consisted in breaking squarely in two, by pushing with his head, a 90-pound steel railroad iron used as the top bar of his fence. He knows the mechanism of the latch of the ponderous steel door between his two box stalls, and nothing but a small pin that only human fingers can manipulate suffices to thwart his efforts to control the latch.

Kartoum has gone over every inch of surface of his two apartments, his doors, gates and fences, to find something that he can break or damage. The steel linings of his apartment walls, originally five feet high, we have been compelled to extend upward to a height of nine feet, to save the brick walls from being battered and disfigured. He has searched his steel fences throughout, in order to find their weakest points, and concentrate his attacks upon them. If the sharp-pointed iron spikes three inches long that are set all over his doors are perfectly solid, he respects them, but if one is the least bit loose in its socket, he works at it until he finally breaks it off.

I invite any Doubting Thomas who thinks that Kartoum does not "think" and "reason" to try his own thinking and reasoning at inventing for Kartoum's door a latch that a keeper can easily and surely open and close at a distance of ten feet, and that will be Kartoum-proof. As for ourselves, three or four seemingly intelligent officers and keepers, and a capable foreman of construction, have all they can do to keep ahead of that one elephant, so great is his ingenuity in thwarting our ways and means to restrain him.

In about two days of effort our elephant keepers taught Gunda to receive a coin from the hand of a visitor, or pick it off the floor, lift the lid of a high-placed cash-box, drop the coin into it and ring a bell. This very amusing industry was kept up for several years, but finally it became so popular that it had to be discontinued.

Keeper Dick Richards easily taught Alice to blow a mouth organ, and to ring a telephone, to take the receiver off its hook and hold it to her ear and listen. For years Alice has rendered, every summer, valuable services of a serious nature in carrying children and other visitors around her yard, and only once or twice has she shown a contrary or obstinate spirit.

Tame elephants never tread on the feet of their attendants or knock them down by accident; or, at least, no instances of the kind have come to my knowledge. The elephant's feet are large, his range of vision is circumscribed, and his extreme and wholly voluntary solicitude for the safety of his human attendants can not be due to anything else than independent reasoning. The most intelligent dog is apt to greet his master by planting a pair of dirty paws against his coat or trousers. The most sensible carriage-horse is liable to step on his master's foot or crowd him against a wall in a moment of excitement; but even inside the keddah, with wild elephants all about, and a captive elephant hemmed in by three or four tame animals, the noosers safely work under the bodies and between the feet of the tame elephant until the feet of the captive are tied.

All who have witnessed the tying of captives in a keddah wherein a whole wild herd has been entrapped, testify to the uncanny human- like quality of the intelligence displayed by the tame elephants who assist in tying, leading out and subjugating the wild captives. They enter into the business with both spirit and understanding, and as occasion requires will deceitfully cajole or vigorously punish a troublesome captive. Sir Emerson Tennent asserts that the tame elephants display the most perfect conception of every movement, both of the object to be attained and the means to accomplish it.

Memory in the Elephant. So far as memory may be regarded as an index of an animal's mental capacity, the weight of evidence is most convincingly creditable to the elephant. As a test of memory in an animal, we hold that a trained performance surpasses all others. During the past forty years millions of people have witnessed in either Barnum's or Ringling Brothers' shows, or in the two combined, an imitation military drill performed by from twelve to twenty elephants which in animals of any other species would be considered a remarkable performance. The following were the commands given by one trainer, understood and remembered by each elephant, and executed without any visible hesitation or mistake. These we will call the

Accomplishments of Performing Elephants.

1. Fall in line.

2. Roll-call. (As each elephant's name is called, he takes his place in the ranks).

3. Present arms. (The trunk is uplifted, with its tip curved forward and held in that position for a short time.)

4. Forward, march.

5. File left, march.

6. Right about face, march.

7. Left about face, march.

8. Right by twos, march.

9. Double quick, march.

10. Single file, march.

11. File right.

12. Halt.

13. Ground arms. (All lie down, and lie motionless.)

14. Attention (All arise.)

15. Shoulder arms. (All stand up on their hind-legs.)

In all, fifteen commands were obeyed by the whole company of elephants.

It being impossible, or at least impracticable, to supply so large a number of animals with furniture and stage properties for a further universal performance, certain individuals were supplied with the proper articles when necessary for a continuation of the performance, as follows:

16. Ringing bells.

17. Climbing up a step-ladder.

18. Going lame in a fore leg.

19. Going lame in a hind leg.

20. Stepping up on a tub turned bottom up.

[Illustration with caption: TAME ELEPHANTS ASSISTING IN TYING A WILD CAPTIVE The captive elephant is marked "C." The tame elephants have been quietly massed around him to keep him still and to give the noosers a chance to work at his legs from under the bodies of the tame elephants. The black figures on the tame elephants are their mahouts, wrapped in blankets and lying down. (From A. G. R. Theobald, Mysore)]

21. Standing on a tub on right legs only.

22. The same, on opposite legs.

23. The same, on the fore legs only.

24. The same, on the hind legs only.

25. Using a fan.

26. Turning a hand-organ.

27. Using a handkerchief to wipe the eyes.

28. Sitting in a chair.

29. Kneeling, with the fore legs.

30. Kneeling with the hind legs.

31. Walking astride a man lying lengthwise.

32. Stepping over a man lying down.

33. Forming a pyramid of elephants, by using tubs of various sizes.

While it is true that not all of the acts in the latter part of this performance were performed by each one of the elephants who went through the military drill, there is no reason to doubt the entire ability of each individual to be trained to obey the whole thirty-three commands, and to remember them all accurately and without confusion. The most astonishing feature of the performance, aside from the perfect obedience of the huge beasts, was their easy confidence and accuracy of memory.

We come now to a consideration of the Accomplishments of Working Elephants. In all the timber-forests of southern India every captive elephant is taught to perform all the following acts and services, as I have witnessed on many occasions:

1. To *salaam,* or salute, by raising the trunk.

2. To kneel, to receive a load or a passenger.

3. When standing, to hold up a fore-foot, to serve the driver as a step in climbing to his place.

4. To lie down to be washed, first on one side and then on the other.

5. To open the mouth. 6. To "hand up" any article from the ground to the reach of a person riding.

7. To pull down an obstructing bough.

8. To halt.

9. To back.

10. To pick up the end of a drag-rope and place it between the teeth.

11. To drag a timber.

12. To kneel and with the head turn a log over, or turn it with the tusks if any are present.

13. To push a log into position parallel with others.

14. To balance and carry timbers on the tusks, if possessing tusks of sufficient size.

15. To "speak," or trumpet.

16. To work in harness.

Every working elephant in India is supposed to possess the intelligence necessary to the performance of all the acts enumerated above at the command of his driver, either by spoken words, a pressure of the knees or feet, or a touch with the driving goad. For the sake of generalization I have purposely excluded from this list all tricks and accomplishments which are not universally taught to working elephants. We have seen, however, that performing elephants are capable of executing nearly double the number of acts commonly taught to the workers; and, while it is useless to speculate upon the subject, it must be admitted that, were a trainer to test an elephant's memory by ascertaining the exact number of commands it could remember and execute in rotation, the result would far exceed anything yet obtained. For my own part, I believe it would exceed a hundred. The performance in the circus-ring is limited by time and space, and not by the mental capacity of the elephants.

Comprehension under Training. When we come to consider the comparative mental receptivity and comprehension of animals under man's tuition, we find the elephant absolutely unsurpassed. On account of the fact that an elephant is about eighteen years in coming to anything like maturity, according to the Indian Government standard for working animals, it is far more economical and expeditious to catch full-grown elephants in their native jungles, and train them, than it is to breed and rear them. About ninety per cent of all the elephants now living in captivity were caught in a wild state and tamed, and of the remainder at least eighty per cent were born in captivity of females that were gravid when captured. It will be seen, therefore, that the elephant has derived no advantage whatever from ancestral association with man, and has gained nothing from the careful selection and breeding which, all combined, have made the collie dog, the pointer and the setter the wonderfully intelligent animals they are. For many generations the horse has been bred for strength, for speed, or for beauty of form, but the breeding of the dog has been based *chiefly* on his intelligence as a means to an end. *With all his advantages, it is to be doubted whether the comprehensive faculties of the dog, even in the most exceptional individuals of a whole race, are equal to those of the adult wild elephant fresh from the jungle.*

The extreme difficulty of teaching a dog *of mature age* even the simplest thing is so well known that it has passed into a proverb: "It is hard to teach an old dog new tricks." In other words, the conditions *must* be favorable. But what is the case with the elephant? The question shall be answered by G. P. Sanderson. In his "Wild Beasts of India," he says: *"Nor are there any elephants which can not be easily subjugated, whatever their size or age. The largest and oldest elephants are frequently the most easily tamed, as they are less apprehensive than the younger ones."*

Philosophy of the Elephant in Accepting Captivity and Making the Best of It. The most astounding feature in the education of an elephant is the suddenness of his transition from a wild and lawless denizen of the forest to the quiet, plodding, good- tempered, and cheerful beast of draught or burden. I call it astounding, because in comparison with what could *not* be done with other wild animals caught when adult, no other word is adequate to express the difference. The average wild animal caught fully grown is "a terror," and so far as training is concerned, perfectly impossible.

There takes place in the keddah, or pen of capture, a mighty struggle between the giant strength of the captive and the ingenuity of man, ably seconded by a few powerful tame elephants. When he finds his strength utterly overcome by man's intelligence, he yields to the inevitable, and accepts the situation philosophically. Sanderson once had a narrow escape from death while on the back of a tame elephant inside a keddah, attempting to secure a wild female. She fought his elephant long and viciously, with the strength and courage of despair, but finally she was overcome by superior numbers. Although her attack on Sanderson in the keddah was of the most murderous description, he states that her conduct after her defeat was most exemplary, and she never afterward showed any signs of ill-temper.

Mr. Sanderson and an elephant-driver once mounted a full-grown female elephant *on the sixth day after her capture, without even the presence of a tame animal.* Sir Emerson Tennent records an instance wherein an elephant fed from the hand on the first night of its capture, and in a very few days evinced pleasure at being patted on the head. Such instances as the above can be multiplied indefinitely. To what else shall they be attributed than philosophic reasoning on the part of the elephant? The orang-utan and the chimpanzee, so often put forward as his intellectual superior, when captured alive at any other period than that of helpless infancy, are vicious, aggressive, and intractable not only for weeks and months, but for the remainder of their lives. Orangs captured when fully adult exhibit the most tiger-like ferocity, and are wholly intractable.

If dogs are naturally superior to elephants in natural intellect, it should be as easy to tame and educate newly-caught wild dogs or wolves of mature age, as newly-caught elephants. But, so far from this being the case, it is safe to assert that it would be *impossible* to train even the most intelligent company of pointers, setters or collies ever got together to perform the feats accomplished with such promptness and accuracy by all regularly trained work elephants.

The successful training of all elephants up to the required working point is so fully conceded in India that the market value of an animal depends wholly upon its age, sex, build and the presence or absence of good tusks. The animal's education is either sufficient for the buyer, or, if it is not, he knows it can be made so.

Promptness and Accuracy in the Execution of Man's Orders. This is the fourth quality which serves as a key to the mental capacity and mental processes of an animal.

To me the most impressive feature of a performance of elephants in the circus-ring is the fact that every command uttered is obeyed with true military promptness and freedom from hesitation, and so accurately that an entire performance often is conducted and concluded without the repetition of a single command. One by one the orders are executed with the most human-like precision and steadiness, amounting sometimes to actual nonchalance. Human beings of the highest type scarcely could do better. To some savage races — for example, the native Australians, the Veddahs of Ceylon, or the Jackoons of the Malay Peninsula, I believe that such a performance would be impossible, even under training. I do not believe their minds act with sufficient rapidity and accuracy to enable a company of them to go through with such a wholly artificial performance as successfully as the elephants do.

The thoughtful observer does not need to be told that the brain of the ponderous quadruped acts, as far as it goes, with the same rapidity and precision as that of an intelligent man, — and this, too, in a performance that is wholly artificial and acquired. In the performance of Bartholomew's horses, of which I once kept a record in detail, even the most accomplished members of his troupe often had to be commanded again and again before they would obey. A command often was repeated for the fifth or sixth time before the desired result was obtained. I noted particularly that not one of his horses, — which were the most perfectly trained of any ever seen by me, — was an exception to this rule, or performed his tasks with the prompt obedience and self-confidence so noticeable in *each one* of the sixteen Barnum elephants. The horses usually obeyed with tardiness and hesitation, and very often manifested nervousness and uncertainty.

In the mind of the elephant, e. g., *each* elephant, there was no confusion of ideas or lapses of memory, but, on the contrary, the mental grasp on the whole subject was so secure and comprehensive that the animal felt himself the master of the situation.

I have never yet seen a performance of trained dogs which could be considered worthy of serious comparison with the accomplishments of either performing or working elephants. In the matter of native educational capacity the dog can not on any grounds be considered the rival of the elephant. The alleged mental superiority of the dog is based almost wholly upon his powers of independent reasoning and observation as exhibited in a state of almost perfect *freedom.* Until the elephant who has grown to maturity under man's influence, is allowed the dog's freedom to plan and execute, no conclusive comparison between them can be made.

Moral Qualities of the Elephant. Finally, we come to a consideration of the elephant's moral qualities that have a direct bearing upon our subject. In India, excepting the professional "rogue," the elephant bears a spotless reputation for patience, amiability and obedience. The "rogue" is an individual afflicted with either an incorrigible disposition, or else is afflicted with insanity, either temporary or permanent. I know of no instance on record wherein a *normal elephant* with a *healthy mind* has been guilty of unprovoked homicide, or even of attempting it. I have never heard of an elephant in India so much as kicking, striking or otherwise injuring either human beings or other domestic animals. There have been several instances, however, of persons killed by elephants which were temporarily insane, or "*must,*" and also by others permanently insane. In America several persons have been killed in revenge for ill treatment. In Brooklyn a female elephant once killed a civilian who burned her trunk with a lighted cigar. It is the misfortune but not the fault of the elephant that in advanced age or by want of necessary exercise, he is liable to be attacked by *must,* or sexual insanity, during which period he is clearly irresponsible for his acts.

So many men have been killed by elephants in this country that of late years the idea has been steadily gaining ground that elephants are naturally ill-tempered, and vicious to a dangerous extent. Under fair conditions, nothing could be farther from the truth. We have seen that in the hands of the "gentle Hindu" the elephant is safe and reliable, and never attacks man except under the circumstances already stated. In this country, however, many an elephant is at the mercy of quick-tempered and sometimes revengeful showmen, who very often do not understand the temperaments of the animals under their control, and who during the traveling season are rendered perpetually ill-tempered and vindictive by reason of overwork and insufficient sleep. With such masters as these it is no wonder that occasionally an animal rebels, and executes vengeance. In Minneapolis in December an elephant once went on a rampage through the freezing of its ears. I am quite convinced that an elephant could by ill treatment be driven to insanity, and I have no doubt that this has been done many times. Our bad elephant, Gunda, was bad by nature, but finally he became afflicted with sexual insanity, for which there was no cure. When commanded by man, the elephant will tear a criminal limb from limb, or crush him to death with his knees, or go out to battle holding a sword in his trunk. He will, when told to do so, attack his kind with fury and persistence; but in the course of many hours, and even days, spent in watching wild herds, I never yet saw a single individual show any signs of impatience or ill-temper toward his fellows.

It is safe to say that, thus far, not one half the elephant's mental capabilities have been developed, or even understood. It would be of great interest to determine by experiment the full educational capacity of this interesting quadruped. It would be equally interesting to determine the limit of its reasoning powers in applied mechanics. An animal that can turn a hand-organ at the proper speed, or ring a telephone and go through the motions of listening with a receiver, can be taught to push a smoothing-plane invented purposely for him; but whether he would learn of himself to plane the rough surface smooth, and let the smooth ones remain untouched, is an open question.

While it is generally fruitless and unsatisfactory to enter the field of speculation, I can not resist the temptation to assert my belief that an elephant can be taught to read written characters, and also to express some of his own thoughts or states of feeling in writing. It would be a perfectly simple matter to prepare suitable appliances by which the sagacious animal could hold a crayon in his trunk, and mark upon a surface adapted to his convenience. Many an elephant has been taught to make chalk-marks on a blackboard. In Julian's work on "The Nature of Animals," the eleventh chapter of the second book, he describes in detail the wonderful performances of elephants at Rome, all of which he saw. One passage is of peculiar interest to us, and the following has been given as a translation: "...I saw them writing letters on Roman tablets with their trunks, neither looking awry nor turning aside. The hand, however, of the teacher was placed so as to be a guide in the formation of the letters; and, while it was writing, the animal kept its eye fixed down in an accomplished and scholar- like manner."

I can conceive how an elephant may be taught that certain characters represent certain ideas, and that they are capable of intelligent combinations. The system and judgment and patient effort which developed an active, educated, and even refined intellect in Laura Bridgman—deaf, dumb and blind from birth— ought certainly to be able to teach a clear-headed, intelligent elephant to express at least *some* of his thoughts in writing.

I believe it is as much an act of murder to wantonly take the life of a healthy elephant as to kill a native Australian or a Central- African savage. If it is more culpable to kill an ignorant human savage than an elephant, it is also more culpable to kill an elephant than an echinoderm. Many men are both morally and intellectually lower than many quadrupeds, and are, in my opinion, as wholly destitute of that indefinable attribute called soul as all the lower animals commonly are supposed to be.

If an investigator like Dr. Yerkes, and an educator like Dr. Howe, should take it in hand to develop the mind of the elephant to the highest possible extent, their results would be awaited with peculiar interest, and it would be strange if they did not necessitate a revision of the theories now common among those who concede an immortal soul to every member of the human race, even down to the lowest, but deny it to all the animals below man.

Curvature in the Brain of an Elephant. There is curvature of the spine; and there is curvature in the brain. It afflicts the human race, and all other vertebrates are subject to it.

In the Zoological Park we have had, and still have, a persistent case of it in a female Indian elephant now twenty-three years of age, named "Alice." Her mental ailment several times manifested itself in Luna Park, her former home; but when we purchased the animal her former owners carelessly forgot to mention it.

Four days after Alice reached her new temporary home in our Antelope House, and while being marched around the Park for exercise, she heard the strident cry of one of our mountain lions, and immediately turned and bolted.

Young as she was at that time, her two strong and able-bodied keepers, Thuman and Bayreuther, were utterly unable to restrain her. She surged straight forward for the front door of the Reptile House, and into that building she went, with the two keepers literally swinging from her ears.

As the great beast suddenly loomed up above the crowd of sightseers in the quiet building, the crowd screamed and became almost panic-stricken.

Partly by her own volition and partly by encouragement, she circumnavigated the turtle-bank and went out.

Once outside she went where she pleased, and the keepers were quite unable to control her. Half an hour later she again headed for the Reptile House and we knew that she would again try to enter.

In view of the great array of plate glass cases in that building, many of them containing venomous cobras, rattlesnakes, moccasins and bushmasters, we were thoroughly frightened at the prospect of that crazy beast again coming within reach of them.

With our men fighting frantically, and exhausted by their prolonged efforts to control her, Alice again entered the Reptile House. As she attempted to pass into the main hall, — the danger zone, — our men succeeded in chaining her front feet to the two steel posts of the guard rail, set solidly in concrete on each side of the doorway. Alice tried to pull up those posts by their roots, but they held; and there in front of the Crocodile Pool the keepers and I camped for the night. We fed her hay and bread, to keep her partially occupied, and wondered what she would do in the morning when we would attempt to remove her.

Soon after dawn a force of keepers arrived. Chaining the elephant's front feet together so that she could not step more than a foot, we loosed the chains from the two posts and ordered her to come to an "about face," and go out. Instead of doing that she determinedly advanced toward the right, and came within reach of twelve handsome glazed cases of live reptiles that stood on a long table. Frantically the men tried to drive her back. For answer she put her two front feet on the top bar of the steel guard rail and smashed ten feet of it to the floor. Then she began to butt those glass snake cages off their table, one by one.

"*Boom!*" "*Bang!*" "*Crash!*" they went on the floor, one after another. Soon fourteen banded rattlesnakes of junior size were wriggling over the floor. "Smash" went more cases. The Reptile House was in a great uproar. Soon the big wall cases would be reached, and then — I would be obliged to shoot her dead, to avoid a general delivery of poisonous serpents, and big pythons from twenty to twenty-two feet long. The room resounded with our shouts, and the angry trumpeting of Alice.

At last, by vigorous work with the elephant hooks, Alice was turned and headed out of the building. A foot at a time she passed out, then headed toward the bear dens. Midway, we steered her in among some young maple trees, and soon had her front legs chained to one of them. Alice tried to push it over, and came near to doing so.

Then we quickly tied her hind legs together,—and she was all ours. Seeing that all was clear for a fall, we joyously pushed Alice off her feet. She went over, and fell prone upon her side. In three minutes all her feet were securely anchored to trees, and we sat down upon her prostrate body.

At that crowning indignity Alice was the maddest elephant in the world for that day. We gave her food, and the use of her trunk, and left her there twenty-four hours, to think it over. She deserved a vast beating with canes; but we gave her no punishment whatever. It would have served no good purpose.

During the interval we telephoned to Coney Island, and asked Dick Richards, the former keeper of Alice, to come and reason with her. Promptly he came,—and he is still guiding as best he can the checkered destinies of that erring female.

When Alice was unwound and permitted to arise,—with certain limitations as to her progress through the world,—it was evident that she was in a chastened mood. She quietly marched to her quarters at the Antelope House, and there we interned her. But that was not all of Alice. Very soon we had to move her to the completed Elephant House, half a mile away. Keeper Richards said that two or three times she had bolted into buildings at Luna Park; so we prepared to overcome her idiosyncrasies by a combination of force and strategy. I had the men procure a strong rope about one hundred feet long, in the middle of which I had them fix a very nice steel hook, large enough to hook suddenly around a post or a tree.

One end of that rope we tied to the left foot of Charming Alice, and the remainder of the rope was carried out at full length in front of her.

Willingly enough she started from the Antelope House, and Richards led her about three hundred feet. Then she stopped, and disregarding all advice and hooks, started to come about, to return to the Antelope House. Quickly the anchor was hooked around the nearest fence post, and Alice fetched up against a force stronger than herself. She was greatly annoyed, but in a few minutes decided to go on.

Another lap of two hundred feet, and the same act was repeated, without the slightest variation.

This process continued for nearly half a mile. By that time we were opposite the Elk House and Alice had become wild with baffled rage. She tried hard to smash fences and uproot trees.

At last she stood still and refused to move another foot; and then we played our ace of trumps. Near by, twenty laborers were working. Calling all hands, they took hold of that outstretched rope, and heading straight for the new Elephant House started a new tug of war. Every "heave-ho" of that hilarious company meant a three-foot step forward for Gentle Alice,—willy-nilly. As she raged and roared, the men heaved and laughed. A yard at a time they pulled that fatal left foot, into the corral and into the apartment of Alice; and she had to follow it.

Ever since that time, Alice has been permanently under arrest, and confined to her quarters; but within the safe precincts of two steel-bound yards she carries children on her back, and in summer earns her daily bread.

Elephant Mentality in the Jungle. Mr. A. E. Ross, while Commissioner of Forests in Burma, had many interesting experiences with elephants, and he related the following:

A bad-tempered mahout who had been cruel to his work-elephant finally so enraged the animal that it attempted to take revenge. To forestall an accident, the mahout was discharged, and for two years he completely disappeared. After that lapse of time he quietly reappeared, looking for an engagement. As the line of elephants stood at attention at feeding time, with a score of persons in a group before them, the elephant instantly recognized the face of his old enemy, rushed for him, and drove him out of the camp.

An ill-tempered and dangerous elephant, feared by everybody, once had the end of his trunk nearly cut off in an accident. While the animal was frantic with the pain of it, Mr. Ross ordered him to lie down. As the patient lay in quiet submission, he dressed the wound and put the trunk in rude bamboo splints. The elephant wisely aided the amateur elephant doctor until the wound healed; and afterward that once dangerous animal showed dog-like affection for Mr. Ross.

XII

THE MENTAL AND MORAL TRAITS OF BEARS

Considered as a group, the bears of the world are supremely interesting animals. In fact, no group surpasses them save the Order Primates, and it requires the enrollment of all the apes, baboons and monkeys to accomplish it.

From sunrise to sunrise a bear is an animal of original thought and vigorous enterprise. Put a normal bear in any new situation that you please, he will try to make himself master of it. Use any new or strange material that you please, of wood, metal, stone or concrete, and he will cheerfully set out to find its weakest points and destroy it. If one board in a wall happens to be of wood a little softer than its fellows, with wonderful quickness and precision he will locate it. To tear his way out of an ordinary wooden cage he asks nothing better than a good crack or a soft knot as a starting point.

Let him who thinks that all animals are mere machines of heredity and nothing more, take upon himself the task of collecting, yarding, housing and KEEPING a collection of thirty bears from all over the world, representing from ten to fifteen species. In a very short time the believer in bear knowledge by inheritance only, will begin to see evidences of new thought.

In spite of our best calculations, in twenty-two years and a total of about seventy bears, we have had three bear escapes. The species involved were an Indian sloth bear, an American black bear and a Himalayan black bear. The troublesome three laboriously invented processes by which, supported by surpassing acrobatics, they were able to circumvent our overhanging bars. Now, did the mothers of those bears bequeath to them the special knowledge which enabled them to perform the acrobatic mid-air feat of warping themselves over that sharp-pointed overhang barrier? No; because none of their parents ever saw steel cage-work of any kind.

Universal Traits. The traits common to the majority of bear *species* as we see them manifested in captivity are the following:

First, playfulness; second, spasmodic treachery; third, contentment in comfortable captivity; fourth, love of water; fifth, enterprise in the mischievous destruction of things that can be destroyed.

The bears of the world are distributed throughout Asia, Borneo, the heavy forests of Europe, all North America, and the northwestern portion of South America. In view of their wonderfully interesting traits, it is surprising that so few books have been written about them. The variations in bear character and habit are almost as wide as the distribution of the species.

There are four books in English that are wholly devoted to American bears and their doings. These are "The Grizzly Bear" and "The Black Bear," by William H. Wright, of Spokane(Scribner's), "The Grizzly Bear," by Enos A. Mills, and "The Adventures of James Capen Adams." In 1918 Dr. C. Hart Merriam published as No. 41 of "North American Fauna" a "Review of the Grizzly and Brown Bears of North America" (U.S. Govt.). This is a scientific paper of 135 pages, the product of many years of collecting and study, and it recognizes and describes eighty-six species and sub-species of those two groups in North America. The classification is based chiefly upon the skulls of the animals.

It is unfortunate that up to date no bear student with a tireless pen has written The Book of Bears. But let no man rashly assume that he knows "all about bears." While many bears do think and act along certain lines, I am constantly warning my friends, "Beware of the Bear! You never can tell what he will do next." I hasten to state that of all the bears of the world, the "pet" bear is the most dangerous.

A Story of a "Pet" Bear. In one of the cities of Canadaa gentleman greatly interested in animals kept a young bear cub, as a pet; and once more I say—if thine enemy offend thee, present him with a black-bear cub. The bear was kept in a back yard, chained to a post, and after his first birthday that alleged "pet" dominated everything within his circumpolar region.

One day a lady and gentleman called to see the pet, to observe how tame and good-natured it was. The owner took on his arm a basket of tempting apples, and going into the bear's territory proceeded to show how the Black One would eat from his owner's hand.

The bear was given an apple, which was promptly eaten. The owner reached for a second, but instead of accepting it, the bear instantly became a raging demon. He struck Mr. C. a lightning-quick and powerful blow upon his head, ripping his scalp open. With horrible growls and bawling, the beast, standing fully erect, struck again and again at his victim, who threw his arms across his face to save it from being torn to pieces. Fearful blows from the bear's claw-shod paws rained upon Mr. C.'s head, and his scalp was almost torn away. In the melee he fell, and the bear pounced upon him, to kill him.

The visiting gentleman rushed for a club. Meanwhile the lady visitor, rendered frantic by the sight of the bear killing her host, did a very brave but suicidally dangerous thing. She *seized the hindquarters* of the bear, gripping the fur in her bare hands, and actually dragged the animal off its victim! Fortunately at that dangerous juncture the lady's husband rushed up with a club, beat the raging animal as it deserved, and mastered it.

The owner of the bear survived his injuries, and by a great effort the surgeons saved his scalp. A "pet" bear in its second year may become the most dangerous of all wild animals. This is because it *seems* so affectionate and docile, and yet is liable to turn in one second,—and without the slightest warning, —into a deadly enemy.

Scores of times we have seen this quick change in temper take place in bears inhabiting our dens. Four bears will be quietly and peacefully consuming their bread and vegetables when,— "*biff!*" Like a stroke of lightning a hairy right arm shoots out and lands with a terriffic jolt on the head of a peaceful companion. The victim roars,—in surprise, pain and protest, and then a fight is on. The aggressor roars and bawls, and follows up his blow as if to exterminate his perfectly inoffensive cage-mate.

Mean and cruel visitors are fond of starting bear fights by throwing into the cages tempting bits of fruit, or peanuts; and sometimes a peach stone kills a valuable bear by getting jammed in the pyloric orifice of the stomach.

The owners of bears should NEVER allow visitors to throw food to them. Unlimited feeding by visitors will spoil the tempers of the best bears in the world.

Power of Expression in Bears. Next to the apes and monkeys, I regard bears as the most demonstrative of all wild animals. The average bear is proficient in the art of expression. The position of his ears, the pose of his head and neck, the mobility of his lips and his walking or his resting attitudes all tell their story.

To facial and bodily expression the bear adds his voice; and herein he surpasses most other wild animals! According to his mood he whines, he threatens, or warns by loud snorting. He roars with rage, and when in pain he cries, or he bawls and howls. In addition to this he threatens an enemy by snapping his jaws together with a mighty ominous clank, accompanied by a warning nasal whine. An angry bear will at times give a sudden rake with his claws to the ground, or the concrete on which he stands. Now, with all this facility for emotional expression, backed by an alert and many-sided mind, boundless energy and a playful disposition, is it strange that bears are among the most interesting animals in the world?

Bears in Captivity. With but few exceptions the bears of the world are animals with philosophic minds, and excellent reasoning power, though rarely equal to that of the elephant. One striking proof of this is the promptness with which adult animals accept *comfortable* captivity, and settle down in contentment. What we mean by comfortable captivity very shortly will be defined.

No bear should be kept in a cage with stone walls and an uneven floor; nor without a place to climb; and wherein life is a daily chapter of inactive and lonesome discomfort and unhappiness. The old-fashioned bear "pit" is an abomination of desolation, a sink- hole of misery, and all such means of bear torture should be banished from all civilized countries.

He who cannot make bears comfortable, contented and happy should not keep any.

A large collection of bears of many species properly installed may be relied upon to reveal many variations of temperament and mentality, from the sanguine and good-natured stoic to the hysterical demon. Captivity brings out many traits of character that in a wild state are either latent or absent.

Prominent Traits of Prominent Species. After twenty years of daily observation we now know that

The grizzly is the most keen-minded species of all bears.

The big Alaskan brown bears are the least troublesome in captivity.

The polar bear lives behind a mask, and is not to be trusted.

The black bear is the nearest approach to a general average in ursine character.

The European brown bears are best for training and performances.

The Japanese black bear is nervous, cowardly and hysterical; the little Malay sun bear is the most savage and unsatisfactory.

The Lesson of the Polar and Grizzly. The polar bears of the north, and the Rocky Mountain grizzlies, a hundred years ago were bold and aggressive. That was in the days of the weak, small-bore, muzzle-loading rifles, black powder and slow firing. Today all that is changed. All those bears have recognized the fearful deadliness of the long-range, high-power repeating rifle, and the polar and the grizzly flee from man at the first sight of him, fast and far. No grizzly attacks a man unless it has been attacked, or wounded, or cornered, or *thinks* it is cornered. As an exception, Mr. Stefansson observed two or three polar bears who seemed to be quite unacquainted with man, and but little afraid of him.

The great California grizzly is now believed to be totally extinct. The campaign of Mr. J. A. McGuire, Editor of *Outdoor Life* Magazine, to secure laws for the reasonable protection of bears, is wise, timely and thoroughly deserving of success because such laws are now needed. The bag limit on grizzlies this side of Alaska should be one per year, and no trapping of grizzlies should be permitted anywhere.

The big brown bears of Alaska have not yet recognized the true deadliness of man. They have vanquished so many Indians, and injured or killed so many white men that as yet they are unafraid, insolent, aggressive and dangerous. They need to be shot up so thoroughly that they will learn the lesson of the polars and grizzlies, — that man is a dangerous animal, and the only safe course is to run from him at first sight.

Bears Learn the Principles of Wild Life Protection. Ordinarily both the grizzlies and black bears are shy, suspicious and intensely "wild" creatures; and therefore the quickness and thoroughness with which they learn that they are in sanctuary is all the more surprising. The protected bears of the Yellowstone Park for years have been to tourists a source of wonder and delight. The black bears are recklessly trustful, and familiar quite to the utmost limits. The grizzlies are more reserved, but they have done what the blacks have very wisely not done. They have broken the truce of protection, and attacked men on their own ground.

Strange to say, of several attacks made upon camping parties, the most serious and most nearly fatal was that of 1917 upon Ned Frost, the well-known guide of Cody, Wyoming, and his field companion. They were sleeping under their wagon, well wrapped from the cold in heavy blankets and comfortables, and it is to their bedding alone that they owe their lives. They were viciously attacked by a grizzly, dragged about and mauled, and Frost was seriously bitten and clawed. Fortunately the bedding engaged the activities of their assailant sufficiently that the two men finally escaped alive.

How Buffalo Jones Disciplined a Bad Grizzly. The most ridiculous and laughable performance ever put up with a wild grizzly bear as an actor was staged by Col. C. J.("Buffalo") Jones when he was superintendent of the wild animals of the Yellowstone Park. He marked down for punishment a particularly troublesome grizzly that had often raided tourists' camps at a certain spot, to steal food.

Very skilfully he roped that grizzly around one of his hind legs, suspended him from the limb of a tree, and while the disgraced and outraged silver-tip swung to and fro, bawling, cursing, snapping, snorting and wildly clawing at the air, Buffalo Jones whaled it with a bean-pole until he was tired. With commendable forethought Mr. Jones had for that occasion provided a moving-picture camera, and this film always produces roars of laughter.

Now, here is where we guessed wrongly. We supposed that whenever and wherever a well-beaten grizzly was turned loose, the angry animal would attack the lynching party. But not so. When Mr. Jones' chastened grizzly was turned loose, it thought not of reprisals. It wildly fled to the tall timber, plunged into it, and there turned over a new leaf. I once said: "C. J., you ought to shoot some of those grizzlies, and teach all the rest of them to behave themselves."

[Illustration with caption: WILD BEARS QUICKLY RECOGNIZE PROTECTION The truce of the black bears of the Yellowstone Park. The grizzlies are not nearly so trustful. Photographed by Edmund Heller, 1921. (All rights reserved.)]

"I know it!" he responded, "I know it! But Col. Anderson won't let me: He says that if we did, some people would make a great fuss about it; and I suppose they would."

Recently, however, it has been found imperatively necessary to teach the Park grizzlies a few lessons on the sanctity of a sanctuary, and the rights of man.

We will now record a few cases that serve to illustrate the mental traits of bears.

Case I. The Steel Panel. Two huge male Alaskan brown bears, Ivan and Admiral, lived in adjoining yards. The partition between them consisted of panels of steel. The upper panels were of heavy bar iron. The bottom panels, each four feet high and six feet long, were of flat steel bars woven into a basket pattern. The ends of these flat bars had been passed through narrow slots in the heavy steel frame, and firmly clinched. We would have said that no land animal smaller than an elephant could pull out one of those panels.

By some strange aberration in management, one day it chanced that Admiral's grizzly bear wife was introduced for a brief space into Ivan's den. Immediately Admiral went into a rage, on the ground that his constitutional rights had been infringed. At once he set to work to recover his stolen companion. He began to test those partition panels, one by one. Finally he found the one that seemed to him least powerful, and he at once set to work to tear it out of its frame.

The keepers knew that he could not succeed; but he thought differently. Hooking his short but very powerful claws into the meshes he braced backward and pulled. After a fierce struggle an upper corner yielded. Then the other corner yielded; and at last the whole upper line gave way.

I reached the scene just as he finished tearing both ends free. I saw him bend the steel panel inward, crush it down with his thousand pounds of weight, and dash through the yawning hole into his rival's arena.

Then ensued a great battle. The two huge bears rose high on their hind legs, fiercely struck out with their front paws, and fought mouth to mouth, always aiming to grip the throat. They bit each other's cheeks but no serious injuries were inflicted, and very soon by the vigorous use of pick-handles the two bear keepers drove the fighters apart.

Case 2. Ivan's Begging Scheme. Ivan came from Alaska when a small cub and he has long been the star boarder at the Bear Dens. He is the most good-natured bear that we have, and he has many thoughts. Having observed the high arm motion that a keeper makes in throwing loaves of bread over the top of the nine-foot cage work, Ivan adopted that motion as part of his sign language when food is in sight outside. He stands up high, like a man, and with his left arm he motions, just as the keepers do. Again and again he waves his mighty arm, coaxingly, suggestively, and it says as plain as print: "Come on! Throw it in! Throw it!"

If there is too much delay in the response, he motions with his right paw, also, both arms working together. It is irresistible. At least 500 times has he thus appealed, and he will do it whenever a loaf of bread is held up as the price of an exhibition of his sign language. Of course Ivan thought this out himself, and put it into practice for a very definite purpose.

Case 3. Ivan's Invention for Cracking Beef Bones. Ivan invented a scheme for cracking large beef bones, to get at the ultimate morsels of marrow. He stands erect on his hind feet, first holds the picked bone against his breast, then with his right paw he poises it very carefully upon the back of his left paw. When it is well balanced he flings it about ten feet straight up into the air. When it falls upon the concrete floor a sufficient number of times it breaks, and Ivan gets his well-earned reward. This same plan was pursued by Billy, another Alaskan brown bear. Case 4. A Bear's Ingenious Use of a Door. When Admiral is annoyed and chased disagreeably by either of his two cage-mates he runs into his sleeping-den, slams the steel door shut from the inside, and thus holds his tormentors completely at bay until it suits him to roll the door back again and come out. At night in winter when he goes to bed he almost always shuts the door tightly from within, and keeps it closed all night. He does not believe in sleeping- porches, nor wide-open windows in sleeping-quarters.

Case 5. Admiral Will Not Tolerate White Boots. Recently our bear keepers have found that Admiral has violent objections to boots of white rubber. Keeper Schmidt purchased a pair, to take the place of his old black ones, but when he first wore them into the den for washing the floor the bear flew at him so quickly and so savagely that he had all he could do to make a safe exit. A second trial having resulted similarly, he gave the boots a coat of black paint. But one coat was not wholly satisfactory to Admiral. He saw the hated white through the one coat of black, promptly registered "disapproval," and the patient keeper was forced to add another coat of black. After that the new boots were approved.

Case 6. The Mystery of Death. Once upon a time we had a Japanese black bear named Jappie, quartered in a den with a Himalayan black bear, — the species with long, black side-whiskers and a white tip to its chin. The Japanese bear was about one-third smaller than the Himalayan black.

One night the Japanese bear died, and in the morning the keepers found it lying on the level concrete top of the sleeping dens.

At once they went in to remove the body; but the Himalayan black bear angrily refused to permit them to touch it. For half an hour the men made one effort after another to coax, or entice or to drive the guardian bear away from the dead body, but in vain. When I reached the strange and uncanny scene, the guardian bear was in a great rage. It took a position across the limp body, and from that it fiercely refused to move or to be driven. As an experiment we threw in a lot of leaves, and the guardian promptly raked them over the dead one and stood pat.

We procured a long pole, and from a safe place on the top of the nearest overhang, a keeper tried to prod or push away the guardian of the dead. The living one snarled, roared, and with savage vigor bit the end of the pole. By the time the bear was finally enticed with food down to the front of the den, and the body removed, seven hours had elapsed.

Now, what were the ideas and emotions of the bear? One man can answer about as well as another. We think that the living bear realized that something terrible had happened to its cage-mate, — in whom he never before had manifested any guardianship interest, — and he felt called upon to defend a friend who was very much down and out. It was the first time that he had encountered the great mystery, Death; and whatever it was, he resented it.

Case 7. A Terrible Punishment. Once we had a particularly mean and vicious young Adirondack black bear named Tommy. In a short time he became known as Tommy the Terror. We put him into a big yard with Big Ben, from Florida, and two other bears smaller than Ben, but larger than himself.

In a short time the Terror had whipped and thoroughly cowed Bruno and Jappie. Next he tackled Ben; but Ben's great bulk was too much for him. Finally he devoted a lot of time to bullying and reviling *through the bars* a big but good-natured cinnamon bear, named Bob, who lived in the next den. In all his life up to that time, Bob had had only one fight. Tommy's treatment of Bob was so irritating to everybody that it was much remarked upon; and presently we learned how Bob felt about it.

One morning while doing the cage work, the keeper walked through the partition gate from Bob's den into Tommy's. He slammed the iron gate behind him, as usual, but this time the latch did not catch as usual. In a moment Bob became aware of this unstable condition. Very innocently he sauntered up to the gate, pushed it open, and walked through into the next den. The keeper was then twenty feet away, but a warning cry from without set him in motion to stop the intruder.

[Illustration with caption: ALASKAN BROWN BEAR "IVAN" BEGGING FOR FOOD He invented the very expressive sign language that he employs.]

[Illustration with caption: THE MYSTERY OF DEATH. Himalayan bear jealously guarding the body of a dead cage-mate.]

Having no club to face, Bob quietly ignored the keeper's broom. Paying not the slightest attention to the three inoffensive bears, Bob fixed his gaze on the Terror, at the far end of the den, then made straight for him. Tommy made a feeble attempt at defense, but Bob seized him by the back, bit him, and savagely shook him as a terrier shakes a rat. The Terror yelled lustily "Murder! Murder! Help!" but none of the other bears made a move for his defense. Bob was there to give Tommy the punishment that was due him for his general meanness and his insulting behavior.

The horrified keeper secured his pike-pole, with a stout spike set in the end for defense, and drove the spike into Bob's shoulder. Bob went right on killing the Terror. Again the keeper drove in his goad, and blood flowed freely; but Bob paid not the slightest attention to this severe punishment.

Then the keeper began to beat the cinnamon over the nose; and that made him yield. He gave the Terror a parting shake, let him go, and with a bloody shoulder deliberately walked out of that den and into his own. The punishment of the Terror went to the full limit, and we think all those bears approved it. In a few hours he died of his injuries.

Case 8. The Grizzly Bear and the String. One of the best illustrations I know of the keenness and originality of a wild bear's mind and senses, is found in Mr. W. H. Wright's account of the grizzly bear he did not catch with an elk bait and two set guns, in the Bitter Root Mountains. This story is related in Chapter VI.

Case 9. Silver King's Memory of His Capture. At this moment we have a huge polar bear who refuses to forget that he was captured in the water, in Kane Basin, and who now avoids the water in his swimming pool, almost as much as any burned child dreads fire. Throughout the hottest months of midsummer old Silver King lies on the rock floor of his huge and handsome den, grouching and grumbling, and not more than once a week enjoying a swim in his spacious pool. No other polar bear of ours ever manifested such an aversion for water. The other polar bears who have occupied that same den loved that pool beyond compare, and used to play in its waters for hours at a time. Evidently the chase of Silver King through green arctic water and over ice floes, mile after mile, his final lassoing, and the drag behind a motor boat to the ship were, to old Silver King, a terrible tragedy. Now he regards all deep water as a trap to catch bears, but, strange to relate, the winter's snow and ice seem to renew his interest in his swimming pool. Occasionally he is seen at play in the icy water, and toying with pieces of ice.

Memory in Bears. I think that ordinarily bear memory for human faces and voices is not long. Once I saw Mr. William Lyman Underwood test the memory of a black bear that for eighteen months had been his household pet and daily companion. After a separation of a year, which the bear spent in a public park near Boston, Mr. Underwood approached, alone, close up to the bars of his cage. He spoke to him in the old way, and called him by his old name, but the bear gave absolutely no sign of recognition or remembrance.

How a Wild Grizzly Bear Caches Food. The silver-tip grizzly bear of the Rocky Mountains has a mental trait and a corresponding habit which seems to be unique in bear character. It is the habit of burying food for future use. Once I had a rare opportunity to observe this habit. It was in the Canadian Rockies of British Columbia, in the month of September(1905), while bears were very activism. John M. Phillips and I shot two large white goats, one of which rolled down a steep declivity and out upon the slide- rock, where it was skinned. The flensed body of the other was rolled over the edge of a cliff, and fell on a brushy soil-covered spot about on the same level as the remains of goat No. 1.

The fresh goat remains were promptly discovered by a lusty young grizzly, which ate to satiety from Goat No. 1. With the remains of. Goat No. 2 the grizzly industriously proceeded to establish a cache of meat for future use.

The goat carcass was dragged to a well chosen spot of seclusion on moss-covered earth. On the steep hillside a shallow hole was dug, the whole carcass rolled into it, and then upon it the bear piled nearly a wagon load of fresh earth, moss, and green plants that had been torn up by the roots. Over the highest point of the carcass the mass was twenty-four inches deep. On the ground the cache was elliptical in shape, and its outline measured about seven by nine feet. On the lower side it was four feet high, and on the upper side two feet. The cache was built around two larch saplings, as if to secure their support. On the uphill side of the cache the ground was torn up in a space shaped like a half moon, twenty-eight feet long by nineteen wide.

I regard that cache as a very impressive exhibit of ursine thought, reasoning and conclusion. It showed more fore-thought and provision, and higher purpose in the conservation of food than some human beings ever display, even at their best. The plains Indians and the buffalo hunters were horribly wasteful and improvident. *The impulse of that grizzly was to make good use of every pound of that meat, and to conserve for the future.*

Survival of the Bears. — The bears of North America have survived thirty thousand years after the lions and the sabre-toothed tigers of La Brea perished utterly and disappeared. But there were bears also in those days, as the asphalt pits reveal. Now, why did not all the bears of North America share the fate of the lions and the tigers? It seems reasonable to answer that it was because the bears were wiser, more gifted in the art of self-preservation, and more resourceful in execution. In view of the omnivorous menu of bears, and their appalling dependence upon small things for food, it is to me marvelous that they now maintain themselves with such astounding success.

A grizzly will dig a big and rocky hole three or four feet deep to get one tiny ground-squirrel, a tidbit so small that an adult grizzly could surely eat one hundred of them, like so many plums, at one sitting. A bear will feed on berries under such handicaps that one would not be surprised to see a bear starve to death in a berry-patch.

But almost invariably the wild bear when killed is fairly well fed and prosperous; and I fancy that no one ever found a bear that had died of cold and exposure. The cunning of the black bear in self-preservation surpasses that of all other large mammal species of North America save the wolf, the white-tailed deer and the coyote. In the game of self-preservation I will back that quartet against all the other large land animals of North America.

What Constitutes Comfortable Captivity. It is impossible for any man of good intelligence to work continuously with a wild animal without learning something of its thoughts and its temper.

In our Zoological Park, day by day and hour by hour our people carry into practical effect their knowledge of the psychology of our mammals, birds and reptiles. In view of the work that we have done during the past twenty-one years of the Park's history, we do not need to apologize for claiming to know certain definite things about wild animal minds. It is my belief that nowhere in the world is there in one place so large an aggregation of dangerous beasts, birds and reptiles as ours. And yet accidents to our keepers from them have been exceedingly few, and all have been slight save four.

Twenty-five years ago I endeavored to plan for the Zoological Society the most humane and satisfactory bear dens on earth. Fortunately we knew something about bears, both wild and captive. Never before have we written out the exact motif of those dens, but it is easily told. We endeavored to give each bear the following things:

A very large and luxurious den, open to the sky, and practically on a level with the world;

Perfect sanitation;

A great level playground of smooth concrete;

High, sloping rocks to climb upon when tired of the level floor;

A swimming pool, always full and always clean;

Openwork steel partitions between cages, to promote sociability and cheerfulness;

Plenty of sunlight, but an adequate amount of shade;

Dry and dark sleeping dens with wooden floors, and

Close-up views of all bears for all visitors.

If there are anywhere in the wilds any bears as healthy, happy and as secure in their life tenure as ours, I do not know of them. The wild bear lives in hourly fear of being shot, and of going to bed hungry.

The service of our bear dens is based upon our knowledge of bear psychology. We knew in the beginning that about 97 per cent of our bears would come to us as cubs, or at least when quite young, and we decided to take full advantage of that fact. All our bears save half a dozen all told, have been trained to permit the keepers of the dens to go into their cages, and to *make no fuss about it.* The bears know that when the keepers enter to do the morning housework, or at any other time for any other purpose, they must at once climb up to the gallery, above the sleeping dens, and stay there until the keepers retire. A bear who is slow about going up is sternly ordered to "Go on!" and if he shows any inclination to disobey, a heavy hickory pick-handle is thrown at him with no uncertain hand.

Now, in grooming a herd of bears, a hickory pick-handle leaves no room for argument. If it hits, it hurts. If it does not hit a bear, it strikes the concrete floor or the rocks with a resound and a rebound that frightens the boldest bear almost as much as being hit. So the bear herd wisely climbs up to the first balcony and sits down to wait. No bear ever leaps down to attack a keeper. The distance and the jolt are not pleasant; and whenever a bear grows weary and essays to climb down, he is sternly ordered back. The keepers are forbidden to permit any familiarities on the part of their bears.

All the bears, save one, that have come to us fully grown, and savage, have been managed by other methods, involving shifting cages.

On two occasions only have any of our keepers been badly bitten in our bear dens. Both attacks were due to over-trustfulness of "petted" bears, and to direct disobedience of fixed orders.

From the very beginning I laid down this law for our keepers, and have repeated it from year to year.

"Make no pets of animals large enough to become dangerous. Make every animal understand and admit day by day that you are absolute master, that it has got to obey, and that if it disobeys, or attacks you, *you will kill it!"*

Familiarity with a dangerous wild animal usually breeds contempt and attack.

Timidity is so fatal that none but courageous and determined men should be chosen, *or be permitted,* to take care of dangerous animals.

In every zoological garden heroic deeds are common; and the men take them all as coming in the day's work. Men in positions of control over zoological parks and gardens should recognize it as a solemn duty to provide good salaries for all men who take care of live wild mammals, birds and reptiles. *A man who is in daily danger of getting hurt should not every waking hour of his life be harried and worried by poverty in his home.*

Let me cite one case of real heroism in our bear dens, which went in with "the day's work," as many others have done. Keeper Fred Schlosser thought it would be safe to take our official photographer, Mr. E. R. Sanborn, into the den of a European brown bear mother, to get a close-up photograph of her and her cubs. Schlosser felt sure that Brownie was "all right," and that he could prevent any accident.

But near the end of the work the mother bear drove her cubs into their sleeping den and then made a sudden, vicious and most unexpected attack upon Keeper Schlosser. She rushed him, knocked him down, seized him by his thigh, bit him severely, and then actually began *to drag him* to the door of her sleeping den! (Just *why* she did this I cannot explain!)

Heroically ignoring the great risk to himself, and thinking of nothing but saving Schlosser, Mr. Sanborn seized the club that had fallen from the keeper's hand when he fell, rushed up to the enraged bear and beat her over the head so savagely and so skilfully that she was glad to let go of her victim and retreat into her den. Then Mr. Sanborn seized Schlosser, dragged him away from the den, and stood guard over him until help came.

XIII

MENTAL TRAITS OF A FEW RUMINANTS

When we wish to cover with a single word the hoofed and horned "big game" of the world, we say Ruminants. That easy and comprehensive name embraces (1) the Bison and Wild Cattle, (2) the Sheep, Goats, Ibexes and Markhors, (3) the Deer Family and (4) the Antelope Family. These groups must be considered separately, because the variations in mind and temperament are quite well marked; but beyond wisdom in self-preservation, I do not regard the intelligence of wild ruminants as being really great.

Intellectually the ruminants are not as high as the apes and monkeys, bears, wolves, foxes and dogs, the domestic horses and the elephants. They are handicapped by feet that are good for locomotion and defense, but otherwise are almost as helpless as so many jointed sticks. This condition closes to the ruminants the possibility of a long program of activities which the ruminant brain might otherwise develop. The ruminant hoof and leg is well designed for swift and rough travel, for battles with distance, snow, ice, mud and flood, and for a certain amount of fighting, but they are inept for the higher manifestations of brain power.

Because of this unfortunate condition, the study of ruminants in captivity does not yield a great crop of results. The free wild animals are far better subjects, and it is from them that we have derived our best knowledge of ruminant thoughts and ways. It is not possible, however, to set forth here any more than a limited number of representative species.

THE BISON AND WILD CATTLE. The American Bison.—Through the age- long habit of the American bison to live in large herds, and to feel, generation after generation, the sense of personal security that great numbers usually impart, the bison early acquired the reputation of being a stolid or even a stupid animal. Particularly was this the case in the days of the greatest bison destruction, when a still-hunter could get "a stand" on a bunch of buffaloes quietly grazing at the edge of the great mass, and slowly and surely shoot down each animal that attempted to lead that group away from the sound of his rifle.

During that fatal period the state of the buffalo mind was nothing less than a tragedy. "The bunch" would hear a report two hundred yards away, they would see a grazing cow suddenly and mysteriously fall, struggle, kick the air, and presently lie still. The individuals nearest dully wondered what it was all about. Those farthest away looked once only, and went on grazing. If an experienced old cow grew suspicious and wary, and quietly set out to walk away from those mysterious noises, "bang!" said the Mystery once more, and she would be the one to fall. On this murderous plan, a lucky and experienced hunter could kill from twenty to sixty head of buffaloes, mostly cows, on a space of three or four acres. The fatal trouble was that each buffalo felt that the presence of a hundred or a thousand others feeding close by was an insurance of *security* to the individual, and so there was no stampede.

But after all, the bison is not so big a fool as he looks. He can think; and he can *learn.*

In 1886, when we were about to set out for Montana to try to find a few wild buffaloes for the National Museum, before the reckless cowboys could find, kill and waste absolutely the last one, a hilarious friend said:

"Pshaw! You don't need to take any rifles! Just get a rusty old revolver, mount a good, sensible horse, ride right up alongside the lumbering old beasts, and shoot them down at arm's length." We went; but not armed with "a rusty old revolver." We found a few buffaloes, but ye gods! How changed they were from the old days! Although only two short years had elapsed since the terminal slaughter of the hundreds of thousands whose white skeletons then thickly dotted the Missouri-Yellowstone divide, *they had learned fear of man*, and also how to preserve themselves from that dangerous wild beast. They sought the remotest bad lands, they hid in low grounds, they watched sharply during every daylight hour, and whenever a man on horseback was sighted they were off like a bunch of racers, for a long and frantic run straight away from the trouble-maker. Even at a distance of two miles, or as far as they could see a man, they would run from him,—not one mile, or two, but five miles, or seven or eight miles, to another wild and rugged hiding-place.

To kill the buffalo specimens that we needed, three cowboys and the writer worked hard for nearly three months, and it was all that we could do to outwit those man-scared bison, and to get near enough to them to kill what we required. Many a time, when weary from a long chase, I thought with bitter scorn of my friend with the rusty-old-revolver in his mind. No deer, mountain sheep, tiger, bears nor elephants,—all of which I have pursued (and sometimes overtaken!)—were ever more wary or keen in self-preservation than those bison who *at last* had broken out from under the fatal spell of herd security. I am really glad that this strange turn of Fortune's wheel gave me the knowledge of the true scope of the buffalo mind before the last chance had passed.

What did a wild buffalo do when he found himself with a broken leg, and unable to travel, but otherwise sound? Did he go limping about over the landscape, to attract enemies from afar, and be quickly shot by a man or torn to pieces by wolves? Not he! With the keen intelligence of the wounded wild ruminant, he chose the line of least resistance, and on three legs fled downhill. He went on down, and kept going, until he reached the bottom of the biggest and most tortuous coulee in his neighborhood. And then what? Instead of coming to rest in a reposeful little valley a hundred feet wide, he chose the most rugged branch he could find, the one with the steepest and highest banks, and up that dry bed, with many a twist and turn, he painfully limped his way. At last he found himself in a snug and safe ditch, precisely like a front line trench seven feet wide, with perpendicular walls and zig- zagging so persistently that the de'il himself could not find him save by following him up to close quarters, and landing upon his horns. There, without food or water, the wounded animal would stand for many days, — in fact, until hunger would force him back to the valley's crop of grass. His wild remedy was to *keep still,* and give that broken leg its chance to knit and grow strong.

I have seen in buffalo skeletons healed bone fractures that filled us with wonder. One case that we shot was a big and heavy bull whose hip socket had been utterly smashed, femur head and all, by a heavy rifle ball; but the bull had escaped in spite of his wound, and he had nursed it until it had healed in *good working order.* We can testify that he could run as well as any of the bisons in his bunch.

Of course young bisons can be tamed, and to a certain extent educated. "Buffalo" Jones broke a pair of two-year-old bulls to work under a yoke, and pull a light wagon. He tried them with bridles and bits, but the buffaloes refused to work with them. With tight-fitting halters, and the exercise of much-muscle, he was able for a time to make them "gee" and "haw." But not for long. When they outgrew his ability in free-hand drawing, he rigged an upright windlass on each side of his wagon-box, and firmly attached a line to each. When the team was desired to "gee," he deftly wound up the right line on its windlass, and vice versa for "haw."

But even this did not last a great while. The motor control was more tentative than absolute. Once while driving beside a creek on a hot and thirsty day, the super-heated buffaloes suddenly espied the water, twenty feet or so below the road. Without having been bidden they turned toward it, and the windlass failed to stop them. Over the cut bank they went, wagon, man and buffalo bulls, "in one red burial blent." Although they secured their drink, their reputation as draught oxen was shattered beyond repair, and they were cashiered the service.

Elsewhere I have spoken of the bison's temper and temperament.

THE WILD SHEEP.—It takes most newly-captured adult mountain sheep about six months in palatial zoo quarters to get the idea out of their heads that every man who comes near them, even including the man who feeds and waters them, is going to kill them, and that they must rush wildly to and fro before it occurs. But there are exceptions.

At the same time, wild herds soon learn the large difference between slaughter and protection, and thereafter accept man's hay and salt with dignity and persistence. The fine big-horn photographs that have been taken of *wild* sheep herds on public highways just outside of Banff, Alberta, tell their own story more eloquently than words can do. The photograph of wild sheep, after only twenty-seven years of protection, feeding in herds in the main street of Ouray, Colorado, is an object lesson never to be forgotten by any student of wild animal psychology. And can any such student look upon such a picture and say that those animals have not thought to some purpose upon the important question of danger and safety to sheep?

Is there anyone left who still believes the ancient and bizarre legend that mountain sheep rams jump off cliffs and alight upon their horns? I think not. People now know enough about anatomy, and the mental traits of wild sheep, to know that nothing of that kind ever occurred save by a dreadful accident, followed by the death of the sheep. No spinal column was ever made by Nature or developed by man that could endure without breaking a headforemost fall from the top of a cliff to the slide-rock bottom thereof.

In Colorado, in May 1907, the late Judge D. C. Beaman of Denver saw a big-horn ram which was pursued by dogs to the precipitous end of a mountain ridge, take a leap for life into space from top to bottom. The distance straight down was "between twenty and twenty-five feet." The ram went down absolutely upright, with his head fully erect, and his feet well apart. He landed on the slide rock on his feet, broke no bones, promptly recovered himself and dashed away to safety. Judge Beaman declared that "the dogs were afraid to approach even as near as the edge of the cliff at the point from which the sheep leaped off."

John Muir held the opinion that the legend of horn-landing sheep was born of the wild descent of frightened sheep down rocks so steep that they *seemed* perpendicular but were not, and the sheep, after touching here and there in the wild pitch sometimes landed in a heap at the bottom,—quite against their will. To me this has always seemed a reasonable explanation.

The big-horn sheep has one mental trait that its host of ardent admirers little suspect. It does not like pinnacle rocks, nor narrow ledges across perpendicular cliffs, nor dangerous climbing. It does not "leap from crag to crag," either up, down or across. Go where you will in sheep hunting, nine times out of ten you will find your game on perfectly safe ground, from which there is very little danger of falling.

In spirit and purpose the big-horns are great pioneers and explorers. They always want to see what is on the other side of the range. They will sight a range of far distant desert mountains, and to see what is there will travel by night across ten or twenty miles of level desert to find out.

It was in the Pinacate Mountains of northwestern Mexico, on the eastern shore of the head of the Gulf of California, that we made our most interesting observations on wild big-horn sheep. On those black and blasted peaks and plains of lava, where nature was working hard to replant with desert vegetation a vast volcanic area, we found herds of short-haired, undersized big-horn sheep, struggling to hold their own against terrific heat, short food and long thirst. It is a burning shame that since our discovery of those sheep hunters of a dozen different kinds have almost exterminated them.

We saw one band of seventeen sheep, close to Pinacate Peak, all so utterly ignorant of the ways of men that they practically refused to be frightened at our presence and our silent guns. We watched them a long time, forgetful of the flight of time. They were not shrewdly suspicious of danger. They fed, and frolicked, and dozed, as much engrossed in their indolence as if the world contained no dangers for them.

One day Mr. John M. Phillips and I shot two rams, for the Carnegie Museum; and the next morning I had the most remarkable lesson that

I ever learned in mountain sheep psychology.

Early on that November morning Mr. Jeff Milton and I left our chilly lair in a lava ravine, and most foolishly left both our rifles at our camp. Hobbling along on foot we led a pack mule over half a mile of rough and terrible lava to a dead sheep. There we quickly skinned the animal, packed the skin and a horned head upon the upper deck of our mule, and started back to camp, leading our assistant. Half way back we looked westward across an eighth of a mile of rough, black lava, and saw standing on a low point a fine big-horn ram. He stood in a statuesque attitude, facing us, and fixedly gazing at us. He was trying to make out what we were, and to determine why a perfectly good pair of sheep horns should grow out of the back of a sorrel mule! Ethically he had a right to be puzzled.

Mr. Milton and I were greatly annoyed by the absence of our rifles; and he proposed that we should leave the mule where he stood, go back to our camp, get our guns, and kill the sheep. Now, even then I was quite well up on the subject of curiosity in wild animals, and I knew to a minute what to count upon as the standing period of sheep, goat or deer. As gently as possible I informed Milton that *no* sheep would ever stand and look at a sorrel mule for the length of time it would take us to foot it over that lava to camp, and return.

But my companion was optimistic, and even skeptical.

"Maybe he will, now!" he persisted. "Let's try it. I think he may wait for us."

Much against my judgment, and feeling secretly rebellious at the folly of it all, I agreed to his plan,—solely to be "a good sport," and to play his game. But *I* knew that the effort would be futile, as well as exhausting. Jeff tied the mule, for the sheep to contemplate.

We went and got those rifles. We were gone fully twenty minutes. When we again reached the habitat of the mule, *that ram was still there!* Apparently he had not moved a muscle, nor stirred a foot, nor even batted an eye. Talk about curiosity in a wild animal! He was a living statue of it.

He continued to hold his pose on his lava point while we stalked him under cover of a hillock of lava, and shot him,—almost half an hour after we first saw him. He had been overwhelmingly puzzled by the uncanny sight of a pair of curling horns like his own, growing out of the back of a long-eared sorrel mule which he felt had no zoological right to wear them. He did his level best to think it out; he became a museum specimen in consequence, and he has gone down in history as the Curiosity Ram.

Mental Attitude of Captured Big-Horn Sheep. In 1906 an enterprising and irrepressible young man named Will Frakes took the idea into his head that he must catch some mountain sheep alive, and do it alone and single-handed. Presently he located a few *Ovis nelsoni* in the Avawatz Mountains near Death Valley, California. Finding a water hole to which mountain sheep occasionally came at night to drink, he set steel traps around it. One by one he caught five sheep of various ages, but chiefly adults. The story of this interesting performance is told in *Outdoor Life* magazine for March, April and May, 1907.

I am interested in the mental processes of those sheep as they came in close contact with man, and were compelled by force of circumstances to accept captivity. Knowing, as all animal men do, the fierce resistance usually made by adult animals to the transition from freedom to captivity, I was prepared to read that those nervous and fearsome adult sheep fought day by day until they died.

But not so. Those sheep showed clear perceptive faculties and good judgment. They were quick to learn that they were conquered, and with amazing resignation they accepted the new life and its strange conditions. In describing the chase on foot in thick darkness of a big old male mountain sheep with a steel trap fast on his foot, Mr. Frakes says:

"A sheep's token of surrender is to lie down and lie still. Once he 'possums, no matter what you do, or how badly you may hurt him, he will never flinch. And when this sheep ("Old Stonewall") was thrown down by the trap, he evidently thought that he was captured, and lay still for a few minutes before he found out the difference, which gave me time to come up with him.... So I went to camp, got a trap clamp and some sacks, made a kind of sled and dragged him in. It was just midnight when I got him tied down, and just sun-up when I got to camp with him. I fixed him up the best I could, stood him up beside the other big-horn and took their pictures. He looked so "rough and ready" that I named him "Old Stonewall." But for all his proud, defiant bearing he has always been a good sheep, *and never tried to fight me.* Still he can hit quick and hard when he wants to, and I have to keep him tied up all the time to keep him from killing the other bucks."

Now, I know not what conclusion others will draw from the above clear and straightforward recital, but to me it established in *Ovis nelsoni* a reputation for quick thinking, original reasoning and sound conclusions. In an incredibly short period those animals came up to the status of tame animals. The five sheep caught by Mr. Frakes were suddenly confronted by new conditions, such as their ancestors had never even dreamed of meeting; and *all of them reacted in the same way.* That was more than "animal behavior." It was Thought, and Reason!

THE GOATS. White Mountain Goat.—I never have had any opportunity to study at length, in the wilds, the mental traits of the markhors, ibexes, gorals or serows. I have however, enjoyed rare opportunities with the white Rocky Mountain goat, on the summits of the Canadian Rockies as well as in captivity.

Where we were, on the Elk River Mountains of East Kootenay, the goats had little fear of man. They did not know that we were in the group of the world's most savage predatory animals, and we puzzled them. Fourteen of them once leisurely looked down upon us from the edge of a cliff, and silently studied us for a quarter of an hour. An hour later three of them ran through our camp. One morning an old billy calmly lay down to rest himself on the mountain side about 300 feet above our tents. At last, however, he became uneasy, and moved away.

This goat is not a timid and fearsome soul, ready to go into a panic in the presence of danger. The old billy believes that the best defense is a vigorous offense. On the spot where Cranbrook, B. C., now stands, an old billy was caught unawares on an open plain and surrounded by Indians, dogs and horses. In the battle that ensued he so nearly whipped the entire outfit that a squaw rushed wildly to the rescue with a loaded rifle, to enable the Red army to win against the one lone goat.

In those mountains the white goat, grizzly bear, mountain sheep, mule deer and elk all live together, in perfect liaison, and never but once have I heard of the goat getting into a fight with a joint-tenant species. A large silver-tip grizzly rashly attacked a full-grown billy, and managed to inflict upon him mortal injuries. Before he fell, however, the goat countered by driving his little skewer-sharp black horns into the vitals of the grizzly with such judgment and precision that the dead grizzly was found by Mr. A. B. Fenwick quite near the dead goat.

We know that the mountain goat is a good reasoner in certain life- or death matters affecting himself.

He knows no such thing as becoming panic-stricken from surprise or fear. An animal that looks death in the face every hour from sunrise until sunset is not to be upset by trifles. We have seen that if a dog and several men corner a goat on a precipice ledge, and hem him in so that there is no avenue of escape, he does not grow frantic, as any deer or most sheep would do, and plunge off into space to certain death. Not he. He stands quite still, glares indignantly upon his enemies, shakes his head, occasionally grits his teeth or stamps a foot, but otherwise waits. His attitude and his actions say:

"Well, it is your move. What are you going to do next?"

Most captive ruminants struggle frantically when put into crates for shipment. White goats very rarely do so. They recognize the inevitable, and accept it with resignation. Captive antelopes and deer often kill themselves by dashing madly against wire fences, but I never knew a white goat to injure itself on a fence. Many a wild animal has died from fighting its shipping crate; but no wild goat ever did so. A white goat will walk up a forty-five degree plank to the roof of his house, climb all over it, and joyously perch on the peak; but no mountain sheep or deer of ours ever did so. *They are afraid!* Only the Himalayan tahr equals the white goat in climbing in captivity, and it will climb into the lower branches of an oak tree, just for fun.

Of all the ruminant animals I know intimately, the white mountain goat is the philosopher-in-chief. Were it not so, how would it be possible for him to live and thrive, and attain happiness, on the savage and fearsome summits that form his chosen home? We must bear in mind that the big-horn does not dare to risk the haunts and trails of his white rivals. Hear the Cragmaster of the Rockies:

[Illustration with caption: THE STEADY-NERVED AND COURAGEOUS MOUNTAIN GOAT He refused to be stampeded off his ledge by men or dog. Photographed at eight feet by John M. Phillips (1905).]

"On dizzy ledge of mountain wall, above the timber-line I hear the riven slide-rock fall toward the stunted pine. Upon the paths I tread secure no foot dares follow me, For I am master of the crags, and march above the scree."

In other chapters I have referred to the temperament and logic of this animal, the bravest mountaineer of all America.

THE DEER. — In nervous energy the species of the Deer Family vary all the way from the nervous and hysterical barasingha to the sensible and steady American elk that can successfully be driven in harness like a horse. As I look over the deer of all nations I am bound to award the palm for sound common-sense and reasoning power to the elk.

A foolishly nervous deer seldom takes time to display high intelligence. Naturally we dislike men, women, children or wild animals who are always ready to make fools of themselves, stampede, and disfigure the landscape.

The Axis Deer is quietly sensible, — so long as there is no catching to be done. Try to catch one, and the whole herd goes off like a bomb. Many other species are similar. No wild deer could act more absurdly than does the axis, the barasingha and fallow, even after generations have been bred in captivity.

The Malay Sambar Deer of the Zoological Park have one droll trait. The adult bucks bully and browbeat the does, in a rather mild way, so long as their own antlers are on their heads. But when those antlers take their annual drop, "O, times! O, manners! What a change!" The does do not lose a day in flying at them, and taking revenge for past tyranny. They strike the hornless bucks with their front feet, they butt them, and they bite out of them mouthfuls of hair. The bucks do not seem, to know that they can fight without their antlers, and so the tables are completely turned. This continues until the new horns grow out, the velvet dries and is rubbed off,—and then quickly the tables are turned again.

No other deer species of my personal acquaintance has ever equalled the American elk of Wyoming in recognizing man's protection and accepting his help in evil times. It is not only a few wise ones, or a few half-domestic bands, but vast wild herds of thousands that every winter rush to secure man's hay in the Jackson Hole country, south of the Yellowstone Park. No matter how shy they *all* are in the October hunting season, in the bad days of January and February they know that the annual armistice is on, and it means hay for them instead of bullets. They swarm in the level Jackson Valley, around S. N. Leek's famous ranch and others, until you can see a square mile of solid gray-yellow living elk bodies. Mr. Leek once caught about 2,500 head in one photograph, all hungry. They crowd around the hay sleds like hungry horses. In their greatest hunger they attack the ranchmen's haystacks, just as far as the stout and high log fences will permit them to go, and many a kindhearted ranchman has robbed his own haystacks to save the lives of starving and despairing elk.

The Yellowstone Park elk know the annual shooting and feeding seasons just as thoroughly as do the men of Jackson Hole.

Once there was a bold and hardy western man who trained a bunch of elk to dive from a forty-foot high platform into a pool of water. I say that he "trained" them, because it really was that. The animals quickly learned that the plunge did nothing more than to shock and wet them, and so they submitted to the part they had to play, with commendable resignation. Some deer would have fought the program every step of the way, and soon worn themselves out; but elk, and also horses, learn that the diving performance is all in the day's work; which to me seems like good logic. A few persons believe that such performances are cruel to the animals concerned, but the diving alone is not necessarily so.

Some deer have far too much curiosity, too much desire to know "What is that?" and "What is it all about?" The startled mule deer leaps out, jumps a hundred feet or more at a great pace, then foolishly stops and looks back, to gratify his curiosity. That is the hunter's chance; and that fatal desire for accurate information has been an important contributory cause to the extermination of the mule deer, or Rocky Mountain "black-tail," throughout large areas. In the Yellowstone Park the once-wild herds of mule deer have grown so tame under thirty years of protection that they completely overrun the parade ground, the officers' quarters, and even enter porches and kitchens for food.

Several authors have remarked upon the habits of the elephant, llama and guanaco in returning to the same spot; and this reminds me of a coincidence in my experience that few persons will believe when I relate it.

In the wild and weird bad-lands of Hell Creek, Montana, I once went out deer hunting in company with the original old hermit wolf-hunter of that region, named Max Sieber. With deep feeling Max told me of a remarkable miss that he had made the previous year in firing at a fine mule deer buck from the top of a small butte; for which I gave him my sympathy.

In the course of our morning's tramp through the very bad-lands that were once the ancestral home of the giant carnivorous dinosaur, yclept *Tyrannosaurus rex,* we won our way to the foot of a long naked butte. Then Sieber said, very kindly:

"If you will climb with me up to the top of this butte I will show you where I missed that big buck."

It was not an alluring proposition, and I thought things that I did not speak. However, being an Easy Mark, I said cheerfully, "All right, Max. Go ahead and show me."

We toiled up to a much-too-distant point on the rounded summit, and as Max slowed up and peered down the farther side, he pointed and began to speak.

"He was standing right down there on that little patch of bare — why!" he exclaimed. "*There's a dee-er there now!* But it's a doe! Get down! Get down!" and he crouched. Then I woke up and became interested.

"It is *not* a doe, Max. I see horns!" — Bang!

And in another five seconds a fine buck lay dead on the very spot where Sieber's loved and lost buck had stood one year previously. But that was only an unbelievable coincidence, — unbelievable to all save old Max.

The natural impulse of the mule deer of those bad-lands when flushed by a hunter is to *run over a ridge,* and escape over the top; but that is bad judgement and often proves fatal. It would be wiser for them to run *down,* to the bottoms of those gashed and tortuous gullies, and escape by zig-zagging along the dry stream beds.

The White-Tailed, or Virginia Deer is the wisest member of the Deer Family in North America, and it will be our last big-game species to become extinct. It has reduced self-preservation to an exact science.

In areas of absolute protection it becomes very bold, and breeds rapidly. Around our bungalow in the wilds of Putman County, New York, the deer come and stamp under our windows, tramp through our garden, feed in broad daylight with our neighbor's cattle, and jauntily jump across the roads almost anywhere. They are beautiful objects, in those wild wooded landscapes of lake and hill.

But in the Adirondacks, what a change! If you are keen you may see a few deer in the closed season, but to see in the hunting season a buck with good horns you must be a real hunter. As a skulker and hider, and a detector of hunters, I know no deer equal to the white-tail. In making a safe get-away when found, I will back a buck of this species against all other deer on earth. He has no fatal curiosity. He will not halt and pose for a bullet in order to have a look at you. What the startled buck wants is more space and more green bushes between the Man and himself.

The Moose is a weird-looking and uncanny monster, but he knows one line of strategy that is startling in its logic. Often when a bull moose is fleeing from a long stern chase, — always through wooded country, — he will turn aside, swing a wide semicircle backward, and then lie down for a rest close up to leeward of his trail. There he lies motionless and waits for man-made noises, or man scent; and when he senses either sign of his pursuer, he silently moves away in a new direction.

The Antelopes of the Old World. The antelopes, gazelles, gnus and hartebeests of Africa and Asia almost without exception live in herds, some of them very large. Owing to this fact their minds are as little developed, individually, as the minds of herd animals generally are. The herd animal, relying as it does upon its leaders, and the security that large numbers always seem to afford, is a creature of few independent ideas. It is not like the deer, elk, sheep or goat that has learned things in the hard school of solitude, danger and adversity, with no one on whom to rely for safety save itself. The basic intelligence of the average herd animal can be summed up in one line:

"Post your sentinels, then follow your leader."

Judging from what hunters in Africa have told me, the hunting of most kinds of African antelopes is rather easy and quiet long- range rifle work. In comparison with any sheep, goat, ibex, markhor and even deer hunting, it must be rather mild sport. A level grassy plain with more or less bushes and small trees for use in stalking is a tame scenario beside mountains and heavy forests, and it seems to me that this sameness and tameness of habitat naturally fails to stimulate the mental development of the wild habitants. In captivity, excepting the keen kongoni, or Coke hartebeest, and a few others, the old-world antelopes are mentally rather dull animals. They seem to have few thoughts, and seldom use what they have; but when attacked or wounded the roan antelope is hard to finish. In captivity their chief exercise consists in rubbing and wearing down their horns on the iron bars of their indoor cages, but I must give one of our brindled gnus extra credit for the enterprise and thoroughness that he displayed in wrecking a powerfully-built watertrough, composed of concrete and porcelain. The job was as well done as if it had been the work of a big-horn ram showing off. But that was the only exhibition of its kind by an African antelope.

The Alleged "Charge" of the Rhinoceros. For half a century African hunters wrote of the assaults of African rhinoceroses on caravans and hunting parties; and those accounts actually established for that animal a reputation for pugnacity. Of late years, however, the evil intentions of the rhinoceros have been questioned by several hunters. Finally Col. Theodore Roosevelt firmly declared his belief that the usual supposed "charge" of the rhinoceros is nothing more nor less than a movement to draw nearer to the strange man-object, on account of naturally poor vision, to see what men look like. In fact, I think that most American sportsmen who have hunted in Africa now share that view, and credit the rhino with very rarely running at a hunter or a party in order to do damage.

The Okapi, of Central Africa, inhabits dense jungles of arboreal vegetation and they are so expert in detecting the presence of man and in escaping from him that thus far, so far as we are aware, no white man has ever shot one! The native hunters take them only in pitfalls or in nooses. Mr. Herbert Lang, of the American Museum of Natural History, diligently hunted the okapi, with native aid, but in spite of all his skill in woodcraft the cunning of the okapi was so great, and the brushy woods were so great a handicap to him, that he never shot even one specimen.

In skill in self-preservation the African bongo antelope seems to be a strong rival of the okapi, but it has been killed by a few white men, of whom Captain Kermit Roosevelt is one.

XIV

MENTAL TRAITS OF A FEW RODENTS

Out of the vast mass of the great order of the gnawing animals of the world it is possible here to consider only half a dozen types. However, these will serve to blaze a trail into the midst of the grand army.

The White-Footed Mouse, or Deer Mouse. On the wind-swept divides and coulees of the short-grass region of what once were the

Buffalo Plains of Montana, only the boldest and most resourceful wild mice can survive. There in 1886 we found a white-footed mouse species (*Peromyscus leucopus*), nesting in the brain cavities of bleaching buffalo skulls, on divides as bare and smooth as golf links.

In 1902, while hunting mule deer with Laton A. Huffman in the wildest and most picturesque bad-lands of central Montana, we pitched our tent near the upper waterhole of Hell Creek. [Footnote: A few months later, acting upon the information of our fossil discoveries that we conveyed to Professor Henry Fairfield Osborn, an expedition from the American Museum of Natural History ushered into the scientific world the now famous Hell Creek fossil bed, and found, about five hundred feet from the ashes of our camp-fire, the remains of *Tyrannosaurus rex*.]For the benefit of our camp-fire, our cook proceeded to hitch his rope around a dry cottonwood log and snake it close up to our tent. When it was cut up, we found snugly housed in the hollow, a nest, made chiefly of feathers, containing five white-footed mice. Packed close against the nest was a pint and a half of fine, clean seed, like radish seed, from some weed of the Pulse Family. While the food-store was being examined, and finally deposited in a pile upon the bare ground near the tent door, the five mice escaped into the sage-brush. Near by stood an old-fashioned buggy, which now becomes a valuable piece of stage property.

The next morning, when Mr. Huffman lifted the cushion of his buggy-seat, and opened the top of the shallow box underneath, the five mice, with their heads close together in a droll-looking group, looked out at him in surprise and curiosity, and at first without attempting to run away. But very soon it became our turn to be surprised.

We found that these industrious little creatures had gathered up every particle of their nest, and every seed of their winter store, and carried all of it up into the seat of that buggy! The nest had been carefully re-made, and the seed placed close by, as before. Considering the number of journeys that must have been necessary to carry all those materials over the ground, plus a climb up to the buggy seat, the industry and agility of the mice were amazing.

By way of experiment, we again removed the nest, and while the mice once more took to the sage-brush, we collected all the seed, and poured it in a pile upon the ground, as before. During the following night, those indomitable little creatures *again* carried nest and seed back into the buggy seat, just as before! Then we gathered up the entire family of mice with their nest and seed, and transported them to New York.

Now, the reasoning of those wonderful little creatures, in the face of new conditions, was perfectly obvious, (1) Finding themselves suddenly deprived of their winter home and store of food, (2) they scattered and fled for personal safety into the tall grass and sagebrush. (3) At night they assembled for a council at the ruins of their domicile and granary. (4) They decided that they must in all haste find a new home, close by, because (5) at all hazards their store of food must be saved, to avert starvation. (6) They explored the region around the tent and camp-fire, and (7) finally, as a last resort, they ventured to climb up the thills of the buggy. (8) After a full exploration of it they found that the box under the seat afforded the best winter shelter they had found. (9) At once they decided that it would do, and without a moment's delay or hesitation the whole party of five set to work carrying those seeds up the thills—a fearsome venture for a mouse—and (10) there before daybreak they deposited the entire lot of seeds. (11) Finding that a little time remained, they carried up the whole of their nest materials, made up the nest anew, and settled down within it for better or for worse.

Now, this is no effort of our imagination. It is a story of actual facts, all of which can be proven by three competent witnesses. How many human beings similarly dispossessed and robbed of home and stores, act with the same cool judgment, celerity and precision that those five tiny creatures then and there displayed?

The Wood Rat, Pack Rat, or Trading Rat. Although I have met this wonderful creature (*Neotoma*) in various places on its native soil, I will quote from another and perfectly reliable observer a sample narrative of its startling mental traits. At Oak Lodge, east coast of Florida, we lived for a time in the home of a pair of pack rats whose eccentric work was described to me by Mrs. C. F. Latham, as follows:

First they carried a lot of watermelon seeds from the ground floor upstairs, and hid them under a pillow on a bed. Then they took from the kitchen a tablespoonful of cucumber seeds and hid them in the pocket of a vest that hung upstairs on a nail. In one night they removed from box number one, eighty five pieces of bee-hive furniture, and hid them in another box. On the following night they deposited in box number one about two quarts of corn and oats.

Western frontiersmen and others who live in the land of the pack rat relate stories innumerable of the absurd but industrious doings of these eccentric creatures. The ways of the pack rat are so erratic that I find it impossible to figure out by any rules known to me the workings of their minds. Strange to say, they are not fiends and devils of malice and destruction like the brown rat of civilization, and on the whole it seems that the destruction of valuable property is not by any means a part of their plan. They have a passion for moving things. Their vagaries seem to be due chiefly to caprice, and an overwhelming desire to keep exceedingly busy. I think that the animal psychologists have lost much by so completely ignoring these brain-busy animals, and I hope that in the future they will receive the attention they deserve. Why experiment with stupid and nerveless white rats when pack rats are so cheap?

It was in the wonderland that on the map is labeled "Arizona" that I met some astonishing evidences of the defensive reasoning power of the pack rat. In the Sonoran Desert, where for arid reasons the clumps of creosote bushes and salt bushes stand from four to six feet apart, the bare level ground between clumps affords smooth and easy hunting-grounds for coyotes, foxes and badgers, saying nothing of the hawks and owls.

Now, a burrow in sandy ground is often a poor fortress; and the dropping spine-clad joints of the tree choyas long ago suggested better defenses. In many places we saw the entrance of pack rat burrows defended by two bushels of spiny choya joints and sticks arranged in a compact mound-like mass. In view of the virtue in those deadly spines, any predatory mammal or bird would hesitate long before tackling a bushel of solid joints to dig through it to the mouth of a burrow.

Did those little animals collect and place those joints because of their defensive stickers,—with deliberate forethought and intention? Let us see.

In the grounds of the Desert Botanical Laboratory, in November 1907, we found the answer to this question, so plainly spread before us that even the dullest man can not ignore it, nor the most skeptical dispute it. We found some pack rat runways and burrow entrances so elaborately laid out and so well defended by choya joints that we may well call the ensemble a fortress. On the spot I made a very good map of it, which is presented on page 164. [Footnote: From "Camp-Fires on Desert and Lava" (Scribner's) page 304.] The animal that made it was the White-Throated Pack Rat (*Neotoma albigula*). The fortress consisted of several burrow entrances, the roads leading to which were defended by carefully constructed barriers of cactus joints full of spines.

The habitants had chosen to locate their fortress between a large creosote bush and a tree-choya cactus (*Opuntia fulgida*) that grew on bare ground, twelve feet apart. When away from home and in danger, the pack rats evidently fled for safety to one or the other of those outposts. Between them the four entrance holes, then in use, went down into the earth; and there were also four abandoned holes.

Connecting the two outposts,—the creosote bush and the choya,— with the holes that were in daily use there were some much-used runways, as shown on the map; and each side of each runway was barricaded throughout its length with spiny joints of the choya. A few of the joints were old and dry, but the majority were fresh and in full vigor. We estimated that about three hundred cactus joints were in use guarding those runways; and no coyote or fox of my acquaintance, nor eke a dog of any sense, would rashly jump upon that spiny pavement to capture a rat.

[Illustration with caption: FORTRESS OF A PACK-RAT, AT TUCSON DEFENDED BY THE SPINY POINTS OF THE TREE CHOYA (*Opuntia fulgida*)]

Beyond the cactus outpost the main run led straight to the sheltering base of a thick mesquite bush and a palo verde that grew tightly together. This gave an additional ten feet of safe ground, or about twenty-five feet in all.

On our journey to the Pinacate Mountains, northwestern Mexico, we saw about twelve cactus-defended burrows of the pack rat, some of them carefully located in the midst of large stones that rendered digging by predatory animals almost impossible.

The beautiful little Desert Kangaroo Rat (*Dipodomys deserti*) has worked out quite a different system of home protection. It inhabits deserts of loose sand and creosote bushes, where it digs burrows innumerable, always located amid the roots of the bushes, and each one provided with three or four entrances, — or exits, as the occasion may require. Each burrow is a bewildering labyrinth of galleries and tunnels, and in attempting to lay bare an interior the loose sand caved in, and the little sprite that lived there either escaped at a distant point or was lost in the shuffle of sand.

The Gray Squirrel (*Sciurus carolinensis*). — This beautiful and sprightly animal quickly recognizes man's protection and friendship, and meets him half way. Go into the woods, sit still, make a noise like a nut, and if any grays are there very soon you will see them. The friendships between our Park visitors and the Park's wild squirrels are one of the interesting features of our daily life. We have an excellent picture of Mrs. Russell Sage sitting on a park bench with a wild gray squirrel in her lap. I have never seen red or fox squirrels that even approached the confidence of the gray squirrel in the truce with Man, the Destroyer, but no doubt generous treatment would produce in the former the gray squirrel's degree of confidence.

I never knew an observer of the home life of the gray squirrel who was not profoundly impressed by the habit of that animal in burying nuts in the autumn, and digging them up for food in the winter and spring. From my office window I have seen our silver-gray friends come hopping through eight or ten inches of snow, carefully select a spot, then quickly bore a hole down through the snow to Mother Earth, and emerge with a nut. Thousands of people have seen this remarkable performance and I think that the majority of them still ask the question: "*How* does the squirrel know precisely where to dig?" That question cannot be answered until we have learned how to read the squirrel mind.

Small city parks easily become overstocked with gray squirrels that are not adequately fed, and the result is, — complaints of "depredations." Of course hungry and half-starved squirrels will depredate, — on birds' nests, fruit and gardens. My answer to all inquirers for advice in such cases is — *feed the squirrels, adequately, and constantly, on cracked corn and nuts, and send away the surplus squirrels.*

At this time many persons know that the wild animals and birds now living upon the earth are here solely because they have had sufficient sense to devise ways and means by which to survive. The ignorant, the incompetent, the slothful and the unlucky ones have passed from earth and joined the grand army of fossils.

Take the case of the Rocky Mountain Pika, or little chief "hare," of British Columbia and elsewhere. It is not a hare at all, and it is so queer that it occupies a family all alone. I am now concerning myself with *Ochotona princeps*, of the Canadian Rockies. It is very small and weak, but by its wits it lives in a country reeking with hungry bears, wolverines and martens. The pika is so small and so weak that in the open he could not possibly dig down below the grizzly bear's ability to dig.

And what does he do to save himself, and insure the survival of the fittest?

He burrows far down in the slide-rock that falls from the cliffs, where he is protected by a great bed of broken stone so thick that no predatory animal can dig through it and catch him. There in those awful solitudes, enlivened only by the crack and rattle of falling slide-rock, the harsh cry of Clark's nut-cracker and the whistling wind sweeping over the storm-threshed summits and through the stunted cedar, the pika chooses to make his home. Over the slide-rock that protects him, the snows of the long and dreary winter pile up from six to ten feet deep, and lie unbroken for months. And how does the pika survive?

[Illustration with caption: WILD CHIPMUNKS RESPOND TO MAN'S PROTECTION. J. Alden Loring and his wild pets]

[Illustration with caption: AN OPOSSUM FEIGNING DEATH]

When he is awake, *he lives on hay, of his own making!*

In September and October, and up to the arrival of the enveloping snow, he cuts plants of certain kinds to his liking, he places them in little piles atop of rocks or fallen logs where the sun will strike them, and he leaves them there until they dry sufficiently to be stored without mildewing. Mr. Charles L. Smith declared that the pikas know enough to change their little hay piles as the day wears on, from shade to sunlight. The plants to be made into hay are cut at the edge of the slide-rock, usually about a foot in length, and are carried in and placed on flat- topped rocks around the mouth of the burrow. The stems are laid together with fair evenness, and from start to finish the haymaking of the pika is conducted with admirable system and precision. When we saw and examined half a dozen of those curing hay piles, we felt inclined to take off our hats to the thinking mind of that small animal which was making a perfectly successful struggle to hold its own against the winter rigors of the summits, and at the same time escape from its enemies.

The common, every-day Cotton-Tail Rabbit (*Lepus sylvaticus*) is not credited by anyone with being as wise as a fox, but that is due to our own careless habits of thought. It has been man's way, ever since the days of the Cavemen, to underrate all wild animals except himself. We are not going to cite a long line of individual instances to exhibit the mental processes or the natural wisdom of the rabbit. All we need do is to point to its success in maintaining its existence in spite of the enemies arrayed against it.

Take the state of Pennsylvania, and consider this list of the rabbit's mortal enemies:

450,000 well-armed men and boys, regularly licensed and diligently gunning throughout six weeks of the year, and actually killing each year about 3,500,000 rabbits!

200,000 farmers hunting on their own farms, without licenses.

Predatory animals, such as dogs, cats, skunks, foxes and weasels.

Predatory birds: hawks, eagles and owls.

Destructive elements: forest fires, rain, snow and sleet.

Now, is it not a wonder that *any* rabbits remain alive in Pennsylvania? But they are there. They refuse to be exterminated. Half of them annually outwit all their enemies — smart as they are; they avoid death by hunger and cold, and they go on breeding in defiance of wild men, beasts and birds. Is it not wonderful — the mentality of the gray rabbit? Again we say — the wild animal must think or die.

In recognizing man's protection and friendship, the rabbit is as quick on the draw as the gray squirrel. In our Zoological Park where we constantly kill hunting cats in order that our little wild neighbors, the rabbits, squirrels and chipmunks may live, the rabbits live literally in our midst. They hang around the Administration Building, rear and front, as if they owned it; and one evening at sunset I came near stepping out upon a pair that were roosting on the official door-mat on the porch. There are times when they seem annoyed by the passage of automobiles over the service road.

To keep hungry rabbits from barking your young apple trees in midwinter, spend a dollar or two in buying two or three bushels of corn expressly for them.

The sentry system of the Prairie-"Dog" in guarding "towns" is very nearly perfect. A warning chatter quickly sends every "dog" scurrying to the mouth of its hole, ready for the dive to safety far below. No! the prairie-"dog," rattlesnake and burrowing owl emphatically do NOT dwell together in peace and harmony in the burrow of the "dog." The rodent hates both these interloping enemies, and carefully avoids them. The pocket gopher does his migrating and prospecting at night, when his enemies are asleep. The gray squirrel builds for itself a summer nest of leaves. At the real beginning of winter the prairie-"dog" tightly plugs up with moist earth the mouth of his burrow; and he packs it with his nose. The round-tailed muskrat of Florida (*Neofiber alleni*) builds a little platform over the water of the marsh in which it lives, on which it builds its nest high and dry. The Hudsonian red squirrel will bark and scold at a human intruder for half an hour.

In Chapter IV I have already accorded the beaver a place with the most intelligent animals of the world. The books that have been written concerning that species have been amply justified. It is, however, impossible to refuse this important animal a place in any chapter devoted to the mental traits of rodents, and I deem it fitting to record here our latest experience with this remarkable species.

Our Last Beaver Experiment. In the autumn of 1921 we emptied and cleaned out our Beaver Pond. The old house originally built by the beavers in the centre of the pond, was for sanitary reasons entirely removed. Work on the pond was not finished until about October 25; and the beavers had no house.

It seemed to me a physical impossibility for the beavers to begin a new house at that late date and unassisted finish it by the beginning of winter. One beaver had escaped, and for the remaining three such a task would be beyond their powers. I decided to give them a helping hand, provided they would accept it, by providing them with a wooden house, which they might if they chose, entirely surround and snugly cover with mud and sticks.

But would they accept it in a grateful spirit, and utilize it? One cannot always tell what a wild animal will do.

With loose earth a low island with a flat top was built to carry the house. Its top was six inches above high-water mark, and (that would, if accepted) be the floor of the permanent house. A good, practicable tunnel was built to an underwater entrance.

Upon that our men set a square, bottomless house of wood, with walls two feet high, and a low roof sloping four ways. Over all this the men piled in a neat mound a lot of tree branches of kinds suitable for beaver food; and with that we left the situation up to the beavers. The finish of our work was made on October 28.

For a week there were no developments. The beavers made no sign of approval or disapproval. And then things began to happen. On November 5 we saw a beaver carrying a small green branch into the house for *bedding!* That meant that our offering was going to be accepted.

The subsequent chronology of that beaver house is as follows:

Nov. 10. The beavers pulled all our brush away from the house, back to a distance of six or seven feet. The house stood fully exposed.

Nov. 11. They began to pile up mud and sticks against the base of the south wall.

Nov. 15. Mud-building to cover the house was in full progress.

Nov. 17. Much of our brush had been placed in the stock of food wood being stored for winter use in the pond west of the house.

Nov. 29. The outside of the house was completely covered up to the edges of the roof. The beavers were working fast and hard. No freezing weather yet.

Dec. 15. The roof was not yet covered. Ice had formed on the pond, and house-building operations were at an end until the spring of 1922.

XV

THE MENTAL TRAITS OF BIRDS

In comparison with mammalian mentality, the avian mind is much more elementary and primitive. It is as far behind the average of the mammals as the minds of fishes are inferior to those of reptiles.

Instinct Prominent in Birds. The average bird is more a creature of instinct than of reason. Primarily it lives and moves by and through the knowledge that it has inherited, rather than by the observations it has made and the things it has thought out in its own head.

But let it not for one moment be supposed that the instinctive knowledge of the bird is of a mental quality inferior to that of the mammal. The difference is in kind only, not in degree. As a factor in self-preservation the keen and correct reasoning of the farm-land fox is in no sense superior to the wonderful instinct and prescience of the golden plover that, on a certain calendar day, or week, bids farewell to its comfortable breeding-grounds in the cold north beyond the arctic circle, rises high in the air and launches forth on its long and perilous migration flight of 8,000 miles to its winter resort in Argentina.

The Migrations of Birds. Volumes have been written on the migrations of birds. The subject is vast, and inexhaustable. It is perhaps the most wonderful of all the manifestations of avian intelligence. It is of interest chiefly to the birds of the temperate zone, whose summer homes and food supplies are for four months of the year buried under a mantle of snow and ice. All but a corporal's guard of the birds of the United States and Canada must go south every winter or perish from starvation and cold. It is a case of migrate or die. Many of the birds do not mind the cold of the northern winter—if it is dry; and *if they could be fed in winter,* many of them would remain with us throughout the year.

Consider the migratory habits of our own home favorites, and see what they reveal. After all else has been said, bird migration is the one unfathomable wonder of the avian world. Really, we know of it but little more than we know of the songs of the morning stars. We have learned when the birds start; we know that many of them fly far above the earth; we know where some of them land, and the bird calendars show approximately when they will return. And is not that really about *all* that we do know?

[Illustration with caption: MIGRATION OF THE GOLDEN PLOVER From
"Bird Migration,", by Dr. W. W. Cooke, U. S. Department of Agriculture, 1915.]

What courage it must take, to start on the long, tiresome and dangerous journey! How do they know where to go, far into the heart of the South, to find rest, food and security? When and where do they stop on the way to feed? Vast areas are passed over without alighting; for many species never are seen in mid career. Why is it that the golden plover feels that it is worth while to fly from the arctic coast to Argentina?

Let any man—if one there be—who is not profoundly impressed by the combined instinct and the reasoning of migratory birds do himself the favor to procure and study the 47-page pamphlet by Dr. Wells W. Cooke, of the U. S. Department of Agriculture, entitled

"Bird Migration." I wish I could reproduce it entire; but since that is impossible, here are a few facts and figures from it.

The Bobolink summers in the northern United States and southern Canada, and winters in Paraguay, making 5000 miles of travel each way.

The Scarlet Tanager summers in the northeastern quarter of the United States and winters in Colombia, Equador and northern Peru, a limit to limit flight of 3,880 miles.

The Golden Plover (*Charadrius dominicus*).—"In fall it flies over the ocean from Nova Scotia to South America, 2,400 miles—the longest known flight of any bird. In spring it returns by way of the Mississippi Valley. Thus the migration routes form an enormous ellipse, with a minor axis of 2,000 miles and a major axis stretching 8,000 miles from arctic America to Argentina." (Cooke.) The Arctic Tern (*Sterna paradisaea*), is "the champion long-distance migrant of the world. It breeds as far north as it can find land on which to build its nest, and winters as far south as there is open water to furnish it food. The extreme summer and winter homes are 11,000 miles apart, or a yearly round trip of 22,000 miles." (Cooke.)

By what do migrating birds guide their courses high in air on a pitch-dark night,—their busy time for flying? Do they, too, know about the mariner's Southern Cross, and steer by it on starlit nights? Equally strange things have happened.

The regular semi-annual migrations of birds may fairly be regarded as the high-water mark of instinct so profound and far-reaching that it deserves to rank as high as reason. To me it is one of the most marvelous things in Nature's Book of Wonders. I never see a humming-bird poised over a floral tube of a trumpet creeper without pausing, in wonder that is perpetual, and asking the eternal question: "Frail and delicate feathered sprite, that any storm-gust might dash to earth and destroy, and that any enemy might crush, *how* do you make your long and perilous journeys unstarved and unkilled? Is it because you bear a charmed life? What is the unsolved mystery of your tiny existence in this rough and cruel world?"

We understand well enough the foundation principles of mammalian and avian life, and existence under adverse circumstances. The mammal is tied to his environment. He cannot go far from the circumpolar regions of his home. A bear chained to a stake is emblematic of the universal handicap on mammalian life. Survive or perish, the average land-going quadruped must stay put, and make the best of the home in which he is born. If he attempts to migrate fast and far, he is reasonably certain to get into grave danger, and lose his life.

The bird, however, is a free moral agent. If the purple grackle does not like the sunflower seeds in my garden, lo! he is up and away across the Sound to Oyster Bay, Long Island, where his luck may be better. Failing there, he gives himself a transfer to Wilmington, or Richmond, via his own Atlantic coast line.

The wonderful migratory instincts of birds have been developed and intensified through countless generations by the imperative need for instinctive guidance, and the comparatively small temptation to inductive reasoning based on known facts. Evidently the bird is emboldened to migrate by the comfortable belief that somewhere the world contains food and warmth to its liking, and that if it flies fast enough and far enough it will find it.

As a weather prophet, the prescience of the bird is strictly limited. The warm spells of late February deceive the birds just as they do the flowers of the peach tree and the apple. Often the bluebirds and robins migrate northward too early, encounter blizzards, and perish in large numbers from snow, sleet, cold and hunger.

The Homing Sense of Birds. We can go no farther than to say that while the homing instinct of certain species of birds is quite well known, the mental process by which it functions is practically unknown. The direction instinct of the homing pigeon is marvelous, but we know that that instinct does not leap full- fledged from the nest. The homer needs assistance and training. When it is about three months old, it is taken in a basket to a point a mile distant from its home and liberated. If it makes good in returning to the home loft, the distances are increased by easy stages — two, three, five, ten, twenty, thirty, fifty and seventy- five miles usually being flown before the bird is sent as far as 100 miles. The official long-distance record for a homing pigeon is 1689.44 miles, held by an American bird.

The homing instinct, or sense, is present in some mammals, but it is by no means so phenomenal as in some species of birds. In mammals it is individual rather than species-wide. Individual horses, dogs and cats have done wonderful things under the propulsion of the homing instinct, but that instinct is by no means general throughout those species. Among wild animals, exhibitions of the home-finding instinct are rare, but the annals of the Zoological Park contain one amusing record.

For emergency reasons, a dozen fallow deer once were quartered in our Bison range, behind a fence only sixty-six inches high. Presently they leaped out to freedom, disappeared in the thick northern forests of the Bronx, and we charged them up to profit and loss. But those deer soon found that life outside our domain was not the dream of paradise that they had supposed. After about a week of wandering through a cold, unsympathetic and oatless world those were sadder and wiser deer, and one night they all returned and joyously and thankfully jumped back into their range, where they were happy ever after.

Recognition of Sanctuary Protection. In this field of precise observation and reasoning, most birds,—if not indeed all of them,—are quick in discernment and accurate in deduction. The great gauntlet of guns has taught the birds of the United States and Canada to recognize the difference between areas of shooting and no shooting. Dull indeed is the bird mind that does not know enough to return to the feeding-ground in which it has been safe from attack. The wild geese and ducks are very keen about sanctuary waters, and no protected pond or river is too small to command attention. Our own little Lake Agassiz, in the New York Zoological Park, each year is the resort of hundreds of mallards and black ducks. And each year a number of absolutely wild wood ducks breed there and in spite of all dangers rear their young. Our wild-fowl pond, surrounded by various installations for birds, several times has been honored by visiting delegations of wild geese, seven of which were caught in 1902 for exhibition.

The most astounding example of avian recognition of protection and human friendship is the spectacle of Mr. Jack Miner's wild goose sanctuary at Kingsville, Ontario, not far from Detroit. With his tile works on one side and his home on the other, he scooped out between them clay for his factory and made a small pond. With deliberate and praiseworthy intention Mr. Miner planted there a little flock of pinioned wild Canada geese, as a notice of sanctuary and an invitation to wild flocks to come down for food, rest and good society.

Very slowly at first the wild geese began to come; but finally the word was passed along the line from Hudson Bay to Currituck Sound that Miner's roadhouse was a good place at which to stop. Year by year the wild geese came, and saw, and were conquered. So many thousands came that presently Mr. Miner grew tired of spending out of his own pocket more than $700 a year for goose corn; and then the Canadian government most commendably assumed the burden, and made Mr. Miner's farm a national bird preserve. [Footnote: Mr. Miner is writing his wild-goose story into a book: and the story is worth it!]

The annals of wild life protection literature contain many records and illustrations of the remarkable quickness and thoroughness of sanctuary recognition by birds. On the other hand I feel greatly annoyed by the failure of waterfowl to reason equally well regarding the decoys of duck-shooters. They fail to learn, either by experience or hearsay, that small flocks of ducks sitting motionless near a shore are loaded, and liable to go off. They fail to learn that it is most wise to settle well outside such flocks of alleged ducks, and that it is a fatal mistake to plump down on the top of a motionless bunch.

Protective Association of Wasps and Caciques. The colonizing caciques, of South America, representing four genera, are very solicitous of the safety of their colonies. In numerous cases, these colonies are found in association with wasps, one or more nests invariably being found near the nests of the birds. It is natural to infer that this strange association is due to the initiative of the birds. When monkeys attack the birds, the birds need the stinging insects.

As usual in the study of wild creatures, the first thing that we encounter in the wild bird is

Temperament. On this hangs the success or failure of a species in association with man. Temperament in the most intellectual wild creatures is just as evident and negotiable to the human eye as colors are in fur or feathers.

A vastly preponderating number of bird species are of sanguine temperament; and it is this fact alone that renders it possible for us to exhibit continuously from 700 to 800 species of birds. Sensible behavior in captivity is the one conspicuous trait of character in which birds mentally and physically are far better balanced than mammals. But few birds are foolishly nervous or hysterical, and when once settled down the great majority of them are sanguine and philosophical. Birds of a great many species can be caught in an adult state and settled down in captivity without difficulty; whereas all save a few species of mammals, when captured as adults, are irreconcilable fighters and many of them die far too quickly. In a well-regulated zoological park nearly every animal that has been caught when adult is a failure and a nuisance.

To name the species of birds that can be caught fully grown and settled down for exhibition purposes, would create a list of formidable length. It is indeed fortunate for us that this is true; for the rearing of nestlings is a tedious task.

A conspicuous exception to the rule of philosophic sedateness in newly caught birds is the loon, or great northern diver. That bird is so exceedingly nervous and foolish, and so persistent in its evil ways, that never once have we succeeded in inducing a loon to settle down on exhibition and be good. When caught and placed in our kind of captivity, the loon goes daft. It dives and dives, and swims under water until it is completely exhausted; it loses its appetite, and very soon dies. Of course if one had a whole marine biological station to place at the disposal of the foolish loon, it might get on.

There are other odd exceptions to the rule of normal bird conduct. Some of our upland game birds, particularly the Franklin grouse and ptarmigan of the Rocky Mountains, display real mental deficiencies in the very necessary business of self-preservation.

WILDNESS AND TAMENESS OF THE RUFFED GROUSE. The ruffed grouse is one of the most difficult of all North American game birds to keep in captivity. This fact is due largely, though not entirely, to the nervous and often hysterical temperament of this species. Some birds will within a reasonable time quiet down and accept captivity, but others throughout long periods, — or forever, — remain wild as hawks, and perpetually try to dash themselves to pieces against the wire of their enclosures. Prof. A. A. Allen of Cornell once kept a bird for an entire year, only to find it at the end of that time hopelessly wild; so he gave the bird its liberty.

However, in this species there are numerous exceptions. Some wing- tipped birds have calmed down and accepted captivity gracefully and sensibly, and a few of the cases of this kind have been remarkable. The most astonishing cases, however, have been of the tameness of free wild birds, in the Catskills, and also near the city of Schenectady. A great many perfectly truthful stories have been published of wild birds that actually sought close acquaintance with people, and took food from their hands.

We have been asked to account for those strange manifestations, but it is impossible to do so. It seems that in some manner, certain grouse individuals learned that Man is not always a killer and a dangerous animal, and so those birds accepted him as a friend, — until the killers came along and violated the sanctuary status.

It is both necessary, and highly desirable for the increase of species, that all wild birds should fly promptly, rapidly and far from the presence of Man, the Arch Enemy of Wild Life. The species that persistently neglects to do so, or is unable, soon is utterly destroyed. The great auk species was massacred and extirpated on Funk Island because it could not get away from its sordid enemies who destroyed it for a paltry supply of *oil*.

The Fool Hen and Its Folly. In our own country there exists a grouse species so foolish in its mind, and so destitute of the most ordinary instinct of self-preservation that it has been known for many years as "the Fool Hen." Definitely, it is the Franklin Grouse (*Conachites franklini*), and its home is in the foothills of the Rocky Mountains. This famous and pitiable victim of misplaced confidence will sit only eight feet up on a jack pine limb, beside a well travelled road, while Mack Norboe dismounts, finds a suitable stick, and knocks the foolish bird dead from its perch. I have seen these birds sit still and patiently wait for their heads to be shot off, one by one, with a .22 calibre revolver when all points of the compass were open for their escape.

All this, however, must be set down as an unusual and phenomenal absence of the most natural instinct of self-protection. The pinnated grouse, sage grouse, Bob White quail and ptarmigan exercise but little keen reason in self-protection. They are easy marks, — the joy of the pot-hunter and the delight of the duffer "sportsman."

Dullness of Instinct in Grouse and Quail. The pinnated grouse, which in Iowa and the Dakotas positively is a migratory bird, does know enough to fly high when it is migrating, but seemingly this species and the sage grouse never will grow wise enough to save themselves from hunters when on their feeding grounds. In detecting the presence of their arch enemy they are hopelessly dull; and they are slow in taking wing.

The quail is a very good hider, but a mighty poor flyer. When a covey is flushed by a collection of dogs and armed men, the lightning-quick and explosive get-away is all right; but the unshot birds do not fly half far enough! Instead of bowling away for two or three miles and getting clear out of the danger zone and hiding in the nearest timber, what do they do? They foolishly stop on the other side of the field, or in the next acre of brush, in full view of the hunters and dogs, who find it great fun to hustle after them and in fifteen minutes put them up again. Thus it is easy for a hunting party to "follow up" a covey until the last bird of it has been bagged.

Just before the five-year close season on quail went into effect in Iowa, this incident occurred:

On a farm of four hundred acres in the southern part of the state, two gunners killed so nearly up to their bag limit of *fifty birds per day* that in ten days they went away with 400 quail. The foolish birds obstinately refused to leave the farm which had been their home and shelter. Day after day the chase with dogs and men, and the fusillade of shots, went briskly on. As a matter of fact, that outfit easily could have gone on until every quail on that farm had fallen.

It is indeed strange that the very bird which practices such fine and successful strategy in leading an intruder away from its helpless young, by playing wounded, should fail so seriously when before the guns. A hunted quail covey should learn to post a sentry to watch for danger and give the alarm in time for a safe flight.

But I know one quail species that is a glorious exception. It is Gambel's quail, of southern Arizona. I saw a good wing shot, Mr. John M. Phillips, hunt that quail (without dogs) until he was hot and red, and come in with more wrath than birds. He said, with an injured air:

"The little beggars won't rise! I don't want to shoot them on the ground, and the minute they rise above the creosote bushes they drop right down into them again, and go on running."

It was even so. They simply will not rise and fly away, as Bob White does, giving the sportsmen a chance to kill them, but when forced to fly up clear of the bushes they at once drop back again. [Footnote: A very few quail-killers of the East who oppose long close seasons contend that quail coveys "breed better" when they are shot to pieces every year and "scattered," but we observed that the quail of the Sonoran Desert managed to survive and breed and perpetuate themselves numerously without the benevolent cooperation of the "pump-gun" and the automatic shotgun.] While the study of avian mentality is a difficult undertaking, this is no excuse for the fact that up to this date (1922) that field of endeavor has been only scratched on its surface. The birds of the world are by no means so destitute of ideas and inventions that they merit almost universal neglect. Because of the suggestions they contain we will point out a few prominent mental traits in birds, chosen at random.

At the same time, let us all beware of seeing too much, and chary of recording scientific hallucinations. It is better to see nothing than to see many things that are not true! In ten octavo pages that particular rock can split wide open the best reputation ever grown.

Bird Architecture. The wisdom of birds in the selection of nesting sites, the designing of the best nest for their respective wants, and finally the construction of them, indicate instinct, reasoning power and mechanical skill of a high order. The range from the wonderful woven homes of the weaver bird and the Baltimore oriole down to the bare and nestless incubating spot of the penguin is so great that nothing less than a volume can furnish space in which to set it forth. But let us at least take a brief glance at a wide range of home-building activities by birds.

The orioles, caciques and weavers weave wonderful homes of fibrous material, often in populous communities.

The bower birds erect remarkable bowers, as playhouses.

The brush turkey scratches together a huge mound of sticks and leaves, four feet by ten or twelve wide at the base.

The vireo and many others turn out beautiful cup-like nests.

The hummingbird builds with the solidity and tenacity of the wasp.

The swallow is a wonderful modeler with mud.

The guacharo builds a solid nest like a cheese with a concave top.

The auklet, the puffin and the kingfishers burrow into the friendly and solid earth. The eider duck plucks from its own breast the softest, of feather linings for its nest.

[Illustration with caption: REMARKABLE VILLAGE NESTS OF THE SOCIABLE WEAVER BIRD (Copied from "The Fauna of South Africa Birds," by Arthur C. Stark)]

The grebe thoughtfully keeps its nest above high-water mark by building on a floating island.

The murre and the guillemot do their best to escape their enemies of the land by building high upon inaccessible rock ledges.

The woodpecker trusts no living species save his own, and drills high up into a hollow tree-trunk for his home.

The cactus wren and crissal thrasher build in the geographical centres of tree choyas, so protected by 500,000 spines that no hawk or owl can reach them.

This catalogue could be extended to a great length; but why pile evidence upon evidence!

It cannot be correct to assume that the nesting activities of birds are based upon instinct alone. That theory would be untenable.

New conditions call for independent thought, and originality of treatment. If the ancestral plans and specifications could not be varied, then every bird would have to build a nest just "such as mother used to make," or have no brood.

All bird students know full well how easily the robin, the wren, the hawk and the owl change locations and materials to meet new and strange conditions. A robin has been known to build on the running-board of a switch-engine in a freight yard, and another robin built on the frame of the iron gate of an elephant yard. A wren will build in a tin can, a piece of drain tile, a lantern, a bird house or a coat pocket, just as blithely as its grandmother built in a grape arbor over a kitchen door. All this is the hall mark of New Thought.

Whenever children go afield in bird country, they are constantly on the alert for fresh discoveries and surprises in bird architecture. Interest in the nest-building ingenuity and mechanical skill of birds is perpetual. The variety is almost endless. Dull indeed is the mind to which a cunningly contrived nest does not appeal. Tell the boys that it is *all right* to collect *abandoned* nests, but the taking of eggs and occupied nests is unlawful and wicked.

The Play-House of the Bower Bird. Years ago we read of the wonderful playhouses constructed by the bower birds of Australia and New Guinea, but nothing ever brought home to us this remarkable manifestation of bird thought so closely as did the sight of our own satin bower bird busily at work on his own bower. He was quartered in the great indoor flying cage of our largest bird house, and supplied with hard grass stems of the right sort for bower-making.

With those materials, scattered over the sand floor, the bird built his bower by taking each stem in his beak, holding it very firmly and then with a strong sidewise and downward thrust slicking it upright in the sand, to stand and to point "just exactly so." The finished bower was a Gothic tunnel with walls of grass stems, about eighteen inches long and a foot high. In making it the male bird wrought as busily as a child building a playhouse of blocks. Our bird would pick up pieces of blue yarn that had been placed in his cage to test his color sense, but never red, — which color seemed to displease him. As the bird worked quietly yet diligently, one could not help longing to know what thoughts were at work in that busy little brain.

The most elaborate of all the bower bird play-houses is that constructed by the gardener bower bird, which is thus described by Pycraft in his "History of Birds":

"This species builds at the foot of a small tree a kind of hut or cabin, some two feet in height, roofed with orchid stems that slope to the ground, regularly radiating from the central support, which is covered with a conical mass of moss sheltering a gallery round it. One side of this hut is left open, and in front of it is arranged a bed of verdant moss, bedecked with blossoms and berries of the brightest color. As the ornaments wither they are removed to a heap behind the hut and replaced by others that are fresh. The hut is circular and some three feet in diameter, and the mossy lawn in front of it is nearly twice that expanse. Each hut and garden is believed to be the work of a single pair of birds. The use of the hut, it appears, is solely to serve the purpose of a playing-ground, or as a place wherein to pay court to the female, since it, like the bowers built by its near relatives, are built long before the nest is begun, this, by the way, being placed in a tree."

[Illustration with caption: SPOTTED BOWER-BIRD, AT WORK ON ITS
UNFINISHED BOWER Foreground garnished with the bird's playthings.
(From A. S. Le Souef, Sydney. Photo by F. C. Morse)]

Most Birds Fear Man. With the exception of those that have been reared in captivity, nearly all species of wild birds, either in captivity or out of it, fear the touch of man, and shrink from him. The birds of the lawn, the orchard and the farm are always suspicious, always on the defensive. But of course there are exceptions. A naturalist like J. Alden Loring can by patient effort win the confidence of a chickadee, or a phoebe bird, and bring it literally to his finger. These exceptions, however, are rare, but they show conclusively that wild birds can be educated into new ideas.

The shrinking of wild birds from the hand of man is almost as pronounced in captivity as it is in the wilderness, and this fact renders psychological experiments with birds extremely difficult. It is really strange that the parrots and cockatoos all should take kindly to man, trust him and even like him, while nearly all other birds persistently fly, or run, or swim or dive away from him. A bird keeper may keep for twenty years, feeding daily, but his hawks, owls and eagles, the perchers, waders, swimmers and upland game birds all fly from him in nervous fear whenever he attempts to handle them. The exceptions to this rule, out of the 20,000 species of the birds of the world, are few.

Wild Birds that Voluntarily Associate with Man. The species that will do so are not numerous, and I will confine myself to some of those that I have seen.

The Indian adjutant, the mynah, hoopoe, vulture, robin, phoebe bird, bluebird, swallow, barn owl, flicker, oriole, jay, magpie, crow, purple grackle, starling, stork, wood pigeon, Canada goose, mallard, pintail, bob white and a few other species have accepted man at his face value and endeavored to establish with him a modus vivendi. The mallard and the graylag goose are the ancestors of our domestic ducks and geese. The jungle fowls have given us the domestic chickens. The wild turkey, the pheasants, the guinea fowl, the ostrich, the emu and the peacock we possess in domestication unchanged.

Caged Wild Birds Quickly Appreciate Sanctuary. Mr. Crandall reports that in the Zoological Park there have been many instances of the voluntary return to their cages of wild birds that have escaped from them. The following instances are cited, out of many that are remembered:

A wild hermit thrush, only two weeks in captivity, escaped from an outdoor cage. But he refused to leave the vicinity of his new home, and permanent food supply. He lingered around for two or three days, and finally a wise keeper opened the cage door when he was near it, and at once he went in.

A magpie escaped from an outside cage, and for a week he lingered around it unwilling to leave its vicinity. At last the other birds of the cage were removed, the door was left open, and the magpie at once went back home.

Bird Memory and Talk. Birds have few ways and means by which to reveal their powers of memory. The best exhibits are made by the talking parrots and cockatoos. The feats of some of these birds, both in memory and expression, are really wonderful. The startling aptness with which some parrots apply the language they possess often is quite uncanny. Concerning "sound mimicry" and the efforts of memory on which they are based, Mr. Lee S. Crandall, Curator of Birds, has contributed the following statement of his observations:

"Many birds, including practically all members of the parrot tribe, many of the crows and jays, as well as mynas and starlings, learn to repeat sounds, words and sentences. Ability varies with both species and individuals. Certain species show greater aptitude as a whole than other species, while there is a great difference between individuals of the same species. "Gray parrots are generally considered the most intelligent of their tribe, and are especially apt at imitating sounds, such as running water, whistles, etc. I have one at home which always answers a knock with 'Come in.' Often he furnishes the knock himself by pounding the perch with his bill, following it with 'Come in.' Amazon parrots are especially good at tunes, some specimens being able to whistle complicated airs and sometimes sing several verses in a high, clear voice. Both grays and Amazons often talk with great fluency, vocabularies having been reported of as many as one hundred words. Often there seems to be intelligent association of certain acts or conditions with corresponding sentences, these sometimes occurring with singular patness.

"Hill mynahs, of the genus *Eulabes,* often talk as well as parrots. The common introduced European starling often says a few words quite clearly. I once knew a long-tailed glossy starling (*Lamprotornis caudatus*) which shared an aviary with an accomplished albino jackdaw. The starling had acquired much of the jackdaw's repertoire, and the 'conversations' carried on between the two birds were most amusing."

A raven in the Zoological Park says "Arthur," "Shut up," "All out" and "Now look what's here" as perfectly as any parrot.

Listed in the order of their ability to learn and remember talk, the important talking birds are as follows: African gray parrot, yellow-headed Amazon, other Amazons, the hill mynahs, the cockatoos, the macaws, and the various others previously mentioned.

It is safe to assert that all migratory birds display excellent powers of memory, chiefly by returning to their favorite haunts after long absences.

Recognition of Persons. Mr. Crandall says there can be no doubt of the ability of most birds to recognize individual persons. This is seen in the smallest species as well as in the largest. He once saw a bullfinch in the last stages of pneumonia and almost comatose, show an instant reaction to the presence of an owner it had not seen in weeks. Many birds form dislikes for individual persons. This is especially noticeable in the parrot tribe. A large male South American condor was friendly enough with two of his keepers but would instantly attack any other keeper or other person entering his enclosure, whether wearing the uniform or not. With his two approved keepers he was gentleness itself.

Parasitic Nesting Habits. In the bird world there are a few species whose members are determined to get something for nothing, and to avoid all labor in the rearing of their offspring. This bad habit is known of the Old World cuckoos, the American cow- birds, the South American rice grackle (*Cassidix*), and suspected in the pin-tail whydah (*Vidua serena*). It seems to reach its highest point in the cuckoos. It is believed that individuals lay their eggs only in the nests of species whose eggs resemble their own. Apparently much skill and intelligence is required for introducing parasitic eggs at the most favorable moment. This is equally true of other parasites.

Curator Crandall has taken several eggs and young of the rice birds from nests of two species of giant caciques in Costa Rica, but never saw an adult *Cassidix*. It is considered a very rare species, but probably is more sly than scarce. Young cuckoos eject unwelcome nestlings shortly after hatching.

Daily contact with a large and varied collection of birds great and small, gathered from every section of the habitable regions of the earth, naturally produces in time a long series of interesting cases of intelligence and behavior. Out of our total occurrences and observations I will offer two that reveal original thought.

Good Sense of the Wedge-Tailed Eagle. In discussing bird intelligence with Mr. Herbert D. Atkin, keeper of our Eagles Aviary and the cranes and water birds in the Flying Cage, he called to my attention two species of birds which had very much impressed him. Afterward he showed me all that he described. Keeper Atkin regards the wedge-tailed eagle, of Australia, as the wisest species with which he has to deal. In the first place, all four of the birds in that flock recognize the fact that he is a good friend, not an enemy, and each day they receive him in their midst with cheerful confidence and friendship. In the fall when the time comes to catch them, crate them and wheel them half a mile to their winter quarters in the Ostrich House, they do not become frightened, nor fight against being handled, and submit with commendable sense and appreciation.

The one thing on which the wedge-tailed eagle really insists when in his summer quarters, is his daily spray bath from a hose. When his keeper goes in to give the daily morning wash to the cage, the eagles perch close above his head and screech and scream until the spray is turned upon them. Then they spread their wings, to get it thoroughly, and come out thoroughly soaked. When the spray is merely turned upon their log instead of upon the birds as they sit higher up, they fly down and get into the current wherever it may be.

Memory of the Cereopsis Goose. Keeper Atkin also showed me an instance of the wisdom of the cereopsis geese, from Van Diemens Land, South Australia. During the winter those birds are kept in the Wild-Fowl Pond; but in summer they are quartered in a secluded yard of the Crane's Paddock, nearly half a mile away. Twice a year these birds go under their own steam between those two enclosures. When turned out of the Cranes' Paddock last November they at once set out and walked very briskly southward up the Bird's Valley, past the Zebra House. On reaching the Service Road, a quarter of a mile away, they turned to the left and kept on to the Wolf Dens. There they turned to the right and kept on two hundred yards until they reached the walk coming down from the Reptile House. There they turned to the left, crossed the bridge, stopped at the gate to the Wild-Fowl Pond enclosure, and when the gate was opened they entered and declared themselves "at home."

Mr. Atkin says that in spring these birds show just as much interest in going back to their summer home. Falconry. We cannot do otherwise than regard the ancient sport of falconry as a high tribute to the mental powers of the genus *Falco*. The hunting falcons were educated into the sport of hawking, just as a boy is trained by his big brother to shoot quail on the wing. The birds were furnished with hoods and jesses, and other garnitures. They were carried on the hand of the huntsman, and launched at unlucky herons and bitterns as an *intelligent* living force. The hunting falcon entered into the sport like a true sportsman, and he played the game according to the rules. The sport was cruel, but it was politely exciting, and it certainly was a fine exhibition of bird intelligence. Part of that intelligence was instinctive, but the most of it was acquired, by educational methods.

Outstanding Traits in a Few Groups of Birds. In creatures as much lacking in visible expression as most birds are, it is difficult to detect the emotions and temperaments that prevail in the various groups. Only a few can be cited with certain confidence.

Vanity Displays in Birds. The males of a few species of birds have been specially equipped by nature for the display of their natural vanity. Anyone who has seen a Zoological Park peacock working overtime on a Sunday afternoon in summer when the crowds of visitors are greatest, solely to display the ocellated splendor of his tail plumage, surely must conclude that the bird is well aware of the glories of his tail, and also that he positively enjoys showing off to admiring audiences.

These displays are not casual affairs in the ordinary course of the day's doings. It is a common thing for one of our birds to choose a particularly conspicuous spot, preferably on an elevated terrace, from which his display will carry farthest to the eyes of the crowd. Even if the bird were controlled by the will of a trainer for the purpose of vanity display, the exhibition could not possibly be more perfect. Like a good speaker on a rostrum, the bird faces first in one direction and then in another, and occasionally with a slow and stately movement it completely revolves on its axis for the benefit of those in the rear. "Vain as a peacock" is by no means an unjustifiable comparison.

Plumage displays are indulged in by turkeys, the blue bird of paradise, the greater and lesser birds of paradise, the sage grouse and pinnated grouse, ruffed grouse, golden pheasant and argus pheasant.

On the whole, we may fairly set down vanity as one of the well defined emotions in certain birds, and probably possessed by the males in many species which have not been provided by nature with the means to display it conspicuously.

Materials for Study. In seeking means by which to study the mental and temperamental traits of wild birds and mammals, the definite and clearly cut manifestations are so few in kind that we are glad to seize upon everything available. Of the visible evidences, pugnacity and the fighting habit are valuable materials, because they are visible. Much can be learned from the fighting weakness or strength of animals and men.

In our great collections of birds drawn from all the land areas of the globe, our bird men see much fighting. Mr. Crandall has prepared for me in a condensed form an illuminating collection of facts regarding

PUGNACITY IN CAPTIVE BIRDS

1. Most species do more or less competitive fighting for nesting sites or mates, especially:

Gallinaceous birds,—many of which fight furiously for mates;

The Ruff, or Fighting Snipe (*Machetes pugnax*),—very pugnacious for mates;

House Sparrows (*Passer domesticus*) fight for nesting places and mates; and

Some Waterfowl, especially swans and geese, fight for nesting places.

2. Most species which do not depend chiefly upon concealment, fight fiercely in defense of nests or young. Typical examples are:

Geese;

Swans;

The larger Flycatchers;

Birds of prey, especially the more powerful ones, such as Bald Eagles, Duck Hawks and Horned Owls.

3. Some species fight in competition for food. Conspicuous examples are:

The fiercer hawks;

Some carrion eaters, as the King Vulture, Black, Sharp-Shinned, Cooper, Gos and Duck Hawks, which fight in the air over prey.

4. Certain birds show pugnacity in connection with the robber instinct, as:

Bald Eagle, which robs the Osprey;

Skua and Jaeger, which rob gulls.

5. Some species show general pugnacity. Species to be cited are:

Cassowaries, Emus and Ostriches, all of which are more or less dangerous;

Saras Cranes, which strike wickedly and without warning;

Some Herons, especially if confined, and

Birds of Paradise, which are unreasonably quarrelsome.

6. In non-social birds, each male will fight for his own breeding and feeding territory. The struggle for territory is a wide one, and it is now attracting the attention of bird psychologists.

Birds are no more angelic than human beings are. They have their faults and their mean traits, just as we have; but their repertoire is not so great as ours. In every species that we have seen tried out in captivity, the baser passions are present. This is equally true of mammals. In *confinement*, in every herd and in every flock from elephants down to doves, the strong bully and oppress the weak, and drive them to the wall.

The most philosophic and companionable birds are the parrots, parakeets, macaws and cockatoos.

The birds that most quickly recognize protection sanctuaries and accept them, are the geese, ducks and swans.

The game birds most nervous and foolish, and difficult to maintain in captivity, are the grouse, ptarmigan and quail.

The bird utterly destitute of sense in captivity is the loon.

The birds that are most domineering in captivity are the cranes.

The birds that are most treacherous in captivity are the darters (*Anhinga*).

The birds that go easiest and farthest in training are the parrots, macaws and cockatoos.

The most beautiful bird species of the world are about fifty in number; and only a few of them are found among the birds of paradise.

The minds of wild birds are quite as varied and diversified as are the forms and habits of the different orders and genera. XVI

THE WISDOM OF THE SERPENT

OF all the vertebrates, the serpents live under the greatest handicaps. They are hated and destroyed by all men, they can neither run nor fly far away, and they subsist under maximum difficulties. Those of the temperate zone are ill fitted to withstand the rigors of winter.

And yet the serpents survive; and we have not heard of any species having become extinct during our own times.

It is indeed worth while to "consider the wisdom of the serpent." Without the exercise of keen intelligence all the snakes of the cultivated lands of the world long ago would have been exterminated. The success of serpents of all species in meeting new conditions and maintaining their existence in the face of enormous difficulties compels us, as reasoning beings, to accord to them keen intelligence and ratiocination.

The poisonous serpents afford a striking illustration of reason and folly en masse. The total number of venomous species is really great, and their distribution embraces practically the whole of the torrid and temperate zones. They are too numerous for mention here; and their capacity for mischief to man is very great. Our own country has at least eighteen species of poisonous snakes, including the rattlesnakes, the copperhead, moccasin, and coral snakes. All these, however, are remarkably pacific. Without exception they are non-aggressive, and they attack only when they think they are exposed to danger, and must defend themselves or die. Hundreds of thousands, or even millions, of our people have tramped through the woods and slept in the sage-brush and creosote bushes of the rattlesnake, and waded through swamps full of moccasins, with never a bite. In America only about two persons per year are bitten by *wild* rattlesnakes.

Our snakes, and all but a very few of the other poison-snake species of the world, know that *it pays to keep the peace.* Now, what if all snakes were as foolishly aggressive as the hooded and spectacled cobras of India? Let us see.

Those cobra species are man-haters. They love to attack and do damage. They go out of their way to bite people. They crawl into huts and bungalows, especially during the monsoon rains, and they infest thatch roofs. But are they wise, and retiring, like the house-haunting gopher snake of the South?

By no means. The cobra goes around with a chip on his shoulder. In India they kill from 17,000 to 18,000 people annually! And in return, about 117,000 cobras are killed annually. It is a mighty fortunate thing for humanity on the frontier that the other serpents of the world know that it is a good thing to behave themselves, and not bite unnecessarily.

Fighting Its Own Kind. The Indian cobra, (*Naia tripudians*), is an exception to the rule of serpents that forbids fighting in the family. While cobras in captivity usually do live together in a state of vicious and fully-armed neutrality, sometimes they do fight. One of our cobras once attacked a cage-mate two-thirds the size of itself, vanquished it, seized it by the head and swallowed two-thirds of it before the tragedy was discovered. The assailant was compelled to disgorge his prey, but the victim was very dead.

The poison venom of the cobra, rattlesnake, bushmaster and puff adder is a great handicap on the social standing of the entire serpent family. Mankind in general abhors snakes, both in general and particular. The snake not actually known to be venomous usually is suspected of being so. It is only the strongest mental constitution that can permit a snake to go unkilled when the killing opportunity offers. It is just as natural for the lay brother to kill a chicken snake because it looks like a copperhead, or a hog-nosed blowing "viper" because it looks like a rattlesnake, as it is to shy at a gun that "may be loaded."

To American plainsmen, the non-aggressive temper of the rattlesnake is well known, and it is also a positive asset. I never knew one who was nervously afraid while sleeping in the open that snakes would come and crawl into his bed, or mix up with his camp. Of course all frontiersmen kill rattlers, as a sort of bounden duty to society, but I once knew an eastern man to turn loose a rattlesnake that he had photographed, in the observance of his principle never to kill an animal whose picture he had taken. Subsequently it was gravely reported that one of the restive horses of the outfit had "accidentally" killed that rattler by stepping upon it.

A Summary of Poisonous Snakes. There are about 300,000 poisonous snakes in the United States, and 110,000,000 people for them to bite; but more people are bitten by captive snakes than by wild ones.

A fool and his snake are soon parted.

There are 200,000 rattlesnakes in our country, but all of them will let you alone if you will let them alone.

If your police record is clear, you can sleep safely in the sagebrush.

If ever you need to camp in a cave, remember that in warm weather the rattlesnakes are all out hunting, and will not return until the approach of winter.

The largest snakes of the world exist only in the human mind.

The rattlesnake is a world-beater at minding his own business.

Men do far more fighting per capita than any snakes yet discovered.

The road to an understanding of the minds of serpents is long and difficult. Perhaps the best initial line of approach is through a well-stocked Reptile House. Having studied somewhat in that school I have emerged with a fixed belief that of all vertebrate creatures, snakes are the least understood, and also the most thoroughly misunderstood.

[Illustration: A
PEACE CONFERENCE WITH AN ARIZONA RATTLESNAKE
"You let me alone and
I won't harm you" (From "Camp-Fires on Desert and Lava")]

[Illustration: HAWK-PROOF NEST OF A CACTUS WREN Placed in the centre of a tree choya cactus of Arizona and defended by 10 000 hostile spines (From "Camp-Fires on Desert and Lava")]

The world at large debits serpents with being far more quarrelsome and aggressive than they really are, and it credits them with knowing far less than they do know.

Attitude of Snakes Toward Each Other. Toward each other, the members of the various serpent species are tolerant, patient and peaceful to the last degree. You may place together in one cage twenty big Texas rattlers, or twenty ugly cottonmouth moccasins from the Carolinas, a hundred garter snakes, twenty boa constrictors, or six big pythons, and if the various *species* are kept separate there will be no fighting. You may stir them up to any reasonable extent, and make them keen to strike you, but they do not attack each other.

There are, however, many species that will not mix together in peace. For example, the king snake of New Jersey hates the rattlesnake, no matter what his address may be. Being by habit a constrictor, the king snake at once winds himself tightly around the neck of the rattler, — and proceeds to choke him to death.

The king cobra devours other snakes, as food, and wishes nothing else.

The Gopher Snake. Some snakes that feel sure you will not harm them will permit you to handle them without a protest or a fight. The most spectacular example is the gopher snake of the southeastern United States. This handsome, lustrous, blue-black species is six feet long, shiny, and as clean and smooth as ivory. Its members are famous rat-killers. You can pick up a wild one wherever you find it, and it will not bite you. They do not at all object to being handled, even by timorous lady visitors who never before have touched a live snake; and in the South they are tolerated by farmers for the good they do as rat catchers.

The Wisdom of a Big Python. Once I witnessed an example of snake intelligence on a large scale, which profoundly impressed me.

A reticulated python about twenty-two feet long arrived from Singapore with its old skin dried down upon its body. The snake had been many weeks without a bath, and it had been utterly unable to shed its old skin on schedule time. It was necessary to remove all that dead epidermis, without delay.

The great serpent, fully coiled, was taken out of its box, sprayed with warm water, and gently deposited on the gravel floor of our most spacious python apartment. Later on pails of warm water, sponges and forceps were procured, and five strong keepers were assembled for active service.

The first step was to get the snake safely into the hands of the men, and fully under control. A stream of cold water from a hose was suddenly shot in a deluge upon the python's head, and while it was disconcerted and blinded by the flood, it was seized by the neck, close behind the head. Immediately the waiting keepers seized it by the body, from neck to tail, and straightened it out, to prevent coiling. Strong hands subdued its struggles, and without any violence stretched the writhing wild monster upon the floor.

Then began the sponging and peeling process. The frightened snake writhed and resisted, probably feeling sure that its last hour had come. The men worked quietly, spoke soothingly, and the work proceeded successfully. With the lapse of time the serpent became aware of the fact that it was not to be harmed; for it became quiet, and lay still. At the same time, we all dreaded the crisis that we thought would come when the jaws and the head would be reached.

By the time the head was reached, the snake lay perfectly passive. Beyond all doubt, it understood the game that was being played.

Now, the epidermis of a snake covers the entire head, *including the eyes!* And what would that snake do when the time came to remove the scales from its eyes and lips? It continued to lie perfectly still! When the pulling off of the old skin hurt the new skin underneath, the head flinched slightly, just as any hurt flesh will flinch by reflex action; but that was absolutely all. For a long hour or more, and even when the men pulled the dead scales from those eyes and lips, that strange creature made no resistance or protest. I have seen many people fight their doctors for less.

That wild, newly-caught jungle snake quickly had recognized the situation, and acted its part with a degree of sense and appreciation that was astounding. I do not know of any *adult wild* mammal that would have shown that kind and degree of wisdom under similar circumstances.

Do Snakes "Charm" Birds? Sometimes a wild bird will sit still upon its nest while a big pilot blacksnake, or some other serpent equally bad, climbs up and poises its head before the motionless and terrified bird until at last the serpent seizes the bird to devour it. The bird victim really seems to be "charmed" by its enemy. If there were not some kind of a hypnotic spell cast over the bird, would it not fly away?

I think this strange proceeding is easily explainable by any one with sufficient imagination to put himself in the bird's place. It is the rule of a sitting bird to sit tight, not to be scared off by trifles, and to take great risks rather than expose her eggs to cold and destruction. The ascent and approach of the serpent is absolutely noiseless. Not a leaf is stirred. The potential mother of a brood calmly sits with eyes half closed, at peace with all the world. Suddenly, and with a horrible shock, she discovers a deadly serpent's multi-fanged head and glittering eyes staring at her *within easy striking distance.*

The horrified mother bird feels that she is lost. She knows full well that with any movement to escape the serpent instantly will launch its attack. *Her one hope,* and seemingly her only chance for life, is that *if she remains motionless* the serpent will go its way without harming her. (Think of the thousands of helpless men, women and children who have hoped and acted similarly in the presence of bandits and hold-up men presenting loaded revolvers! But they were far from being "charmed.")

The bird hopes, and sits still, *paralyzed with fear.* At its leisure the serpent strikes; and after a certain number of horrible minutes, all is over. I think there is no real "charm" exercised in the tragedy; but that there is on the part of the bird a paralysis of fear, which is in my opinion a well defined emotion, common in animals and in men. I have seen it in many animals.

Snakes that Feign Death. The common hog-nosed snake, mistakenly called the "puff-adder" and blowing "viper" (*Heterodon platyrhinus*) of the New England states, often feigns death when it is caught in the open, and picked up. It will "play 'possum" while you carry it by its tail, head downward, or hang its limp body over a fence. Of course it hopes to escape by its very clever ruse, and no doubt it often does so from the hands of inexperienced persons.

Do Snakes Swallow Their Young? I *think* not. A number of persons solemnly have declared that they have seen snakes do so, but no *herpetologist* ever has seen an occurrence of that kind. I believe that all of the best authorities on serpents believe that snakes do not swallow their young. The theory of the pro-swallowists is that the mother snake takes her young into her interior to provide for their safety, and that they do not go as far down as the stomach. The anti-swallowists declare that the powerful digestive juices of the stomach of a snake would quickly kill any snakelets coming in contact with it; and I believe that this is true.

At present the snake-swallowing theory must be ticketed "not proven," and is filed for further reference.

The Hoop Snake Fable. There is no such thing as a "hoop-snake" save in the vivid imaginations of a very few men.

The Intelligence of the King Cobra. Curator of Reptiles Raymond L. Ditmars regards the huge king cobra of the Malay Peninsula, the largest of all poisonous serpents, as quite the wisest serpent known to him. He says its mind is alert and responsive to a very unusual degree in serpents, and that it manifests a keen interest in everything that is going on around it, especially at feeding- time. This is quite the reverse of the usual sluggish and apathetic serpent mind in captivity.

Incidentally, I would like very much to know just what our present twelve-foot cobra thought when, upon its arrival at its present home, its total blindness was relieved by the thrillingly skilful removal of the *two layers* of dead scales that had closed over and finally adhered to each orbit.

The vision of the king cobra is keen, and its temper is not easily ruffled. Its temperament seems to be sanguine, which is just the opposite of the nervous-combative hooded and spectacled cobra species.

The So-called "Snake Charmers" of India. Herpetologists generally discredit the idea that a peripatetic Hindu can "charm" a cobra any farther or more quickly than any snake-keeper. In the first place, the fangs of the serpent are totally removed, — by a very savage and painful process. After that, the unfortunate snake is in no condition to fight or to flee. It seeks only to be let alone, and the musical-pipe business is to impress the mind of the observer.

Serpent Psychology an Unplowed Field. At this date (1922) we know only the rudiments of serpent intelligence and temperament. In the wilds, serpents are most elusive and difficult to determine. In captivity they are passive and undemonstrative. We do not know how much memory they have, they rarely show what they think, and on most subjects we do not know where they stand. But the future will change all this. During the past twenty years the number of herpetologists in the United States has increased about tenfold. It is fairly impossible that serpent psychology should much longer remain unstudied, and unrevealed along the lines of plain commonsense.

The Ways of Crocodiles. The ways of crocodiles are dark and deep; their thoughts are few and far between. Their wisdom is above that of the tortoises and turtles, but below that of the serpents. I have had field experience with four species of crocodilians in the New World and three in the Old. With but slight exceptions they all think alike and act alike.

The great salt-water crocodile of the Malay Peninsula and Borneo is the only real man-eater I ever met. Except under the most provocative circumstances, all the others I have met are practically harmless to man. This includes the Florida species, the Orinoco crocodile, the little one from Cuba, the alligator, the Indian gavial and the Indian crocodile (*C. palustris*).

The salt-water crocodile, that I have seen swimming out in the ocean two miles or more from shore, is in Borneo a voracious man-eater. It skilfully stalks its prey in the murky rivers where Malay and Dyak women and children come down to the village bathing place to dip up water and to bathe. There, unseen in the muddy water, the monster glides up stealthily, seizes his victim by the leg, and holding it tightly backs off into deep water and disappears. The victims are drowned, not bitten to death.

I found in Ceylon that the Indian crocodile is a shameless cannibal, devouring the skinned carcasses of its relatives whenever an opportunity offered.

The Florida crocodile is the shrewdest species of all those I know personally. It has the strange habit of digging out deep and spacious burrows for concealment, in the perpendicular sandy banks of southern Florida rivers where the deep water comes right up to the shore. Starting well under low-water mark, the crock digs in the yielding sand, straight into the bank, a roomy subterranean chamber. In this snug retreat he once was safe from all his enemies,—until the fatal day when his secret was discovered, and revealed to a grasping world. Since that time, the Alligator Joes of Palm Beach and Miami have made a business of personally conducting parties of northern visitors, at $50 per catch, to witness the adventure of catching a nine-foot crocodile alive. The dens are located by probing the sand with long iron rods. A rope noose is set over the den's entrance, and when all is ready, a confederate probes the crocodile out of its den and into the fatal noose.

Today the Florida crocodile is so nearly extinct that it required two years of diligent inquiry to produce one live specimen subject to purchase.

Common Sense in the Common Toad. Last spring, in planting a lot of trees on our lawn, a round tree-hole that stood for several days unoccupied finally accumulated about a dozen toads. Its two feet of straight depth was unscalable, and when finally discovered the toads were tired of their imprisonment. Partly as a test of their common-sense, Mr. George T. Fielding placed a six-inch board in the hole, at an angle of about thirty degrees, but fairly leading out of the trap.

In very quick time the toads recognized the possibilities of the inclined plane and hopped upward to liberty. In the use of this opportunity they showed more wisdom than our mountain sheep manifest concerning the same kind of an improvement designed to enable them to reach the roof of their building. XVII

THE TRAINING OF WILD ANIMALS

Before we enter this chapter let us pause a moment on the threshold, and consider the logic of animal training and performances.

Logic is only another name for reason. Its reverse side is fanaticism; and that way madness lies. It is the duty of every sane man and woman to consider the cold logic of every question affecting the welfare of man and nature. Fanaticism when carried to extremes can become a misdemeanor or a crime. The soft-hearted fanaticism of humanics that saves a brutal murderer, or would-be murderer like Berkman, from the gallows or the chair, and eventually turns him loose to commit more crimes against innocent people, is not only wrong, and wicked, but in aggravated cases it is a *crime* against society.

Just now there is a tiny wave of agitation against all performances of trained wild animals, and the keeping of animals in captivity, on the ground that all this is "cruel" and inhumane. The Jacklondon Society of Boston is working hard to get up steam for this crusade, but thus far with only partial success. Its influence is confined to a very small area.

Now, what is the truth of this matter? Is it true that trained wild animals are cruelly abused in the training, or in compelling them to perform? Is it true that in making animals perform on the stage, or in the circus ring, their rights are wickedly infringed? Is it the duty of the American people to stop all performances by animals? Is it wicked to make wild animals, or cats and dogs, *work* for a living, as men and women do? Is it true that captive animals in zoological parks and gardens are miserable and unhappy, and that all such institutions should be "abolished?" What is truth?

In the first place, there is no sound reasoning or logic in assuming that the persons of animals, tame or wild, are any more sacred than those of men, women and children. We hold that it is no more "cruelty" for an ape or a dog to work in training quarters or on the stage than it is for men, women and young people to work as acrobats, or actors, or to engage in honest toil eight hours per day. Who gave to any warm-blooded animal that consumes food and requires shelter the right to live without work? *No one!* I am sure that no trained bear of my acquaintance ever had to work as hard for his food and shelter as does the average bear out in the wilds. In order to find enough to eat the latter is compelled to hustle hard from dawn till dark. I have seen that the Rocky Mountain grizzly feels forced to dig a big hole three feet deep in hard, rocky ground, to get one tiny ground squirrel the size of a chipmunk, — and weighing only eight or nine ounces. Now, has he anything "on" the performing bear? Decidedly not.

I regard the sentimental Jacklondon idea, that no wild animal should be made to work on the stage or in the show-ring, as illogical and absurd. Human beings who sanely work are much happier per capita than those who do nothing but loaf and grouch. I have worked, horse-hard, throughout all the adult years of my life; and it has been good for me. I know that it is no more wrong or wicked for a horse to work for his living, — of course on a humane basis, — either on the stage or on the street, than it is for a coal-carrier, a foundryman, a farmer, a bookkeeper, a school teacher or a housewife to do the day's work.

The person of a wild animal is no more sacred than is that of a man or woman. A sound whack for an unruly elephant, bear or horse is just as helpful as it is for an unruly boy who needs to be shown that order is heaven's first law.

In the presence of the world's toiling and sweating millions, in the presence of millions of children in the home sweat-shops and factories working their little lives out for their daily crust and a hard bed, what shall we think and say of the good, kind-hearted people who are spending time and energy in crusading against trained animal performances?

The vast majority of performing animals are trained by humane men and women, practicing kindness to the utmost; and they are the last persons in the world who would be willing to have their valuable stock roughly handled, neglected or in any manner cruelly treated.

So far as zoological parks and gardens are concerned, they are no more in need of defense than the Rocky Mountains.

Every large zoological park is a school of wild-animal education and training; and it is literally a continuous performance. Let no one suppose that there is no training of wild beasts save for the circus ring and the vaudeville stage. Of the total number of large and important mammals that come into our zoological parks, the majority of them actually are trained to play becomingly their respective parts. An intractable and obstinate animal soon becomes a nuisance.

The following, named in the order of their importance, are the species whose zoological park training is a matter of necessity: Elephants, bears, apes, hippopotami, rhinoceroses, giraffes, bison, musk-ox, wild sheep, goats and deer, African antelopes, wild swine, and wild horses, asses and zebras. Of large birds the most conspicuous candidates for training in park life are the ostriches, emus, cassowaries, cranes, pelicans, swans, egrets and herons, geese, ducks, pheasants, macaws and cockatoos, curassows, eagles and vultures. Among the reptiles, the best trained are the giant tortoises, the pythons, boas, alligators, crocodiles, iguanas and gopher snakes.

Each one of these species is educated (1) to be peaceful, and not attack their keepers; (2) to not fear their keepers; (3) to do as they are bid about going here or there; (4) to accept and eat the food that is provided for them, and (5) finally, in some cases to "show off" a little when commanded, for the benefit of visitors.

All this training comes in the regular course of our daily work, and there are few animals who do not respond to it. The necessity for training is most imperative with the elephants and bears, for without it the difficulties in the management of those dangerous animals is greatly intensified.

In training an animal to do a particular act not in the routine of his daily life, it is of course necessary to show him clearly and pointedly what is desired. I think that in quickness of perception, and ability to adopt a new idea, the elephants and the great apes are tied for first place. Both are remarkably quick. It seems to me that it required only half a dozen lessons to teach our Indian elephant, Gunda, to take a penny in his trunk, lift the lid of a high-placed box, drop in the coin, then pull a bell-cord and ring a bell. Of course the reward for the first successful performances was lumps of sugar. Within three days this rather interesting special exhibit was working smoothly, and coining money. As a means of working off on the poor animal great numbers of foreign copper coins, and spurious issues of all kinds, it was a great boon to the foreign population of New York. Our erratic elephant Alice was quickly trained by Keeper Richards to blow a mouth organ, to ring a telephone by turning the crank, and to take off the receiver and hold it up to her ear for an imaginary call.

Another keeper, with no previous experience as a trainer, taught a male orang-utan called Rajah to go through a series of performances that are elsewhere described.

Bright and Dull Individuals. Every wild animal species contains the same range of bright and dull individuals that are found in the various races of men. Naturally the animal trainer selects for training only those animals that are of amiable disposition, that mentally are alert, responsive and possessed of good memories. The worst mistakes they make are in taking on and forcing ill-natured and irritable animals, that hate training and performing. Often a trainer persists in retaining an animal that resolutely should be thrown out. Captain Bonavita lost his arm solely because of his fatal persistence in retaining in his group of lions an animal that hated him, and which the trainer well knew was dangerous.

While nearly every wild animal can be taught a few simple tricks, the dull mind soon reaches its constitutional limit. Even among the great apes, conditions are quite the same. One half the orang- utans are of the thin-headed, pin-headed type that is hopeless for stage training. The good ones are the stocky, round-headed, round- faced individuals who have the cephalic index of the statesman or jurist, and a broad and well-rounded dome of thought.

Training for the Ring and the Stage. During his long and successful career as a purveyor of wild animals for all purposes, Carl Hagenbeck had great success in the production of large animal groups trained for stage performances. I came in close touch with his methods and their results. His methods were very simple, and they were founded on kindness and common sense. Mr. Hagenbeck hated whips and punishments. When an animal could not get on without them, it was dropped from the cast. His working theory was that an unwilling animal makes a bad actor.

There is no mystery about the best methods in training animals, wild or domestic. The first thing is to assemble a suitable number of *young* animals, all of which are mentally bright and physically sound. Most adult animals are impracticable, and often impossible, because they are set in their ways. The elephants are monumental exceptions. A large, well-lighted and sunny room is provided; and around it are the individual cages for the student animals. The members of the company are fed wisely and well, kept scrupulously clean, and in all ways made comfortable and contented. When not at their work they are allowed to romp and play together until they are tired of the exercise.

The trainer who has been selected to create a specified group spends practically his entire time with his pupils. He feeds them, and mixes with them daily and hourly. From the beginning he teaches them that *they must obey him, and not fight.* The work of training begins with simple things, and goes on to the complex; and each day the same routine is carried out. To each animal is assigned a certain place in the circle, with a certain tub or platform on which to sit at ease when not acting in the ring. It is exceedingly droll to see a dozen cub lions, tigers, bears and cheetahs sitting decorously on their respective tubs and gravely watching the thirteenth cub who is being labored with by the keeper to bring its ideas and acts into line. The stage properties are many; and they all assist in helping the actors to remember the sequence of their acts, as well as the things to be done. The key that controls the mind of a good animal is the reward idea. Many a really bad animal goes through its share of the performance solely to secure the bit of meat, the lump of sugar or the prized biscuit that never fails to show up at the proper moment.

[Illustration with caption: WORK ELEPHANT DRAGGING A HEWN TIMBER
The most primitive form of elephant harness. The end of the drag rope is held between the teeth of the wise and patient animal
(From A. G. R. Theobald, Mysore)]

The acts to be performed are gone over in the training quarters, innumerable times; and this continues so long that by the time the "group" is ready for the stage, behold! the cubs with which the patient and tireless trainer began have grown so large that to the audience they now seem like adult and savage animals. Those who scoff at the wild animal mind, and say that all this displays nothing but "machines in fur" need to be reminded that this very same line of effort in training and rehearsal is absolutely necessary in the production of every military company, every ballet, and every mass performance on the stage. There is *no* successful performance without training. Boys and girls require the very same sort of handling that the wild animals receive, but the humans do with a little less of it.

The man who flouts a good stage performance by wild animals on the ground that it reveals "no thought," and is only "imitation," is, in my judgment, a very short-sighted student. Maniacs and imbeciles cannot be trained to perform any program fit to be seen. I saw that tried fifty years ago, in "the wild Australian children," who were idiots. *The performer must think, and reason.*

Of the many groups of trained animals that I have seen in performances, my mind goes back first to the one which contained a genuine bear comedian, of the Charlie Chaplin type. It was a Himalayan black bear, with fine side whiskers, and it really seemed to me absolutely certain that the other animals in the group appreciated and enjoyed the fun that comedian made. He pretended to be awkward, and frequently fell off his tub. He was purposely dilatory, and was often the last one to finish. The other animals seemed to be fascinated by his mishaps, and they sat on their tubs and watched him with what looked like genuine amusement. I remember another circle of seated animals who calmly and patiently sat and watched while the trainer labored with a cross and refractory leopard, to overcome its stubbornness, and to make it do its part.

Carl Hagenbeck loved to produce mixed groups of dangerous animals,—lions, tigers, leopards, bears and wolves. One trainer whom I knew was assisted in a highly dangerous group by a noble staghound who habitually kept close to his master, and was said to be ready to attack instantly any animal that might attack the trainer. I never saw a finer bodyguard than that dog.

In 1908 the most astounding animal group ever turned out of the Hagenbeck establishment, or shown on any stage, appeared in London. It consisted of *75 full-grown polar bears!* Now, polar bears, either for the cage or the stage, are bad citizens. Instinctively I always suspect their mental reservations, and for twenty-one years have carefully kept our keepers out of their reach. But Mr. William Hagenbeck, brother of the great Carl, actually trained and performed with a huge *herd* of dangerous polars to the number stated.

In the *Strand* magazine for April, 1908, there is a fine article by Arthur Harold about this group and its production. It says that the bears were obtained when seven or eight months old, in large lots, and all thrown in together. It took a keeper between seven and eight months to educate them out of their savage state,—by contact, kindness, sugar and fruit,—and then they were turned over to the trainer, Mr. Hagenbeck. They were taught to form pyramids, climb ladders, shoot the chutes, ride in pony carriages, draw and ride in sleds, drink from bottles, and work a see-saw. Various individuals did individual tricks. The star performer was Monk, the wrestling bear, who went with his trainer through a fearsome wrestling performance.

Concerning the temperament of that polar bear group Mr. William Hagenbeck said:

"Although I know every animal in the company, have taught each one to recognize me, and have been among many of them for *fifteen years,* I can not now tell by their expressions the moods of the animals. This is one of the characteristics of the polar bear. Their expression remains the same, and it is impossible to detect by watching their faces whether they are pleased or cross. Now in most wild animals, such as the lion, you can tell by the expression of the beast's face and by its actions whether it is in a good temper or not.... The truth is, the polar bear is a most awkward beast to train. In the first place its character is difficult to understand. He is by nature very suspicious, and without the least warning is apt to turn upon his trainer. Among the seventy bears that have been taught to do tricks, *only two* of them are really fond of their work."

In the end, Mr. William Hagenbeck was very nearly killed by one of these polar bears. I was with Carl Hagenbeck a few hours after he received telegraphic news of the tragedy, and his bitterness against those polar bears was boundless. I understood that Monk, the wrestling bear, was the assailant,—which was small cause for wonder. When I saw Mr. Hagenbeck's polar bear show, it gave me shivers of fear. The first two big male polars that we installed at our Park came from that very group, and one of them led us into a dreadful tragedy, with a female bear as the victim.

The So-Called "Trick" Performances. Some psychologists make light of what they call "trick performances," in which the performing animals are guided by signs, or signals, or spoken commands from their trainers. I have never been able to account for this. It is incontestably true that dull and stupid animals can learn little, and perform less. For example, all the training in the world could not suffice to put a pig through a performance that a chimpanzee or orang could master in two weeks. The reason is that the pig has not the brain power that is indispensable. A woodchuck never could become the mental equal of a wood rat (*Neotoma*). A sheep could not hope to rival a horse, either in training or in execution.

Really, *the brain, the memory and reason must enter into every animal performance that amounts to anything worth while.* It is just as sensible to flout soldiers on the drill-ground as to wave aside as of no account a troup of trained lions or sea-lions on the stage. Any animal that can be taught to perform difficult feats, and that delivers the goods in the blinding glare and riot of the circus ring or the stage footlights, is entitled to my profound respect for its powers of mind and nerve.

The Sea-Lion's Repertoire. Long ago trainers recognized in the California sea-lion (*Zalophus*) a good subject for the ring and stage. Its long, supple neck, its lithe body and brilliant nervous energy seemed good for difficult acts. The sea-lion takes very kindly to training, and really delights in its performances. In fact, it enters into its performance with a keen vigor and zest that is pleasing to behold. Let this veracious record of a performance of Treat's five sea-lions and two harbor seals, that I witnessed October 15, 1910, tell the whole story, in order that the reader may judge for himself:

1. Each sea-lion balanced upright on its nose a wooden staff 3 feet long, with a round knob on its upper end.

2. Each sea-lion caught in its mouth a three-foot stick with a ball on each end, tossed it up, whirled it in the air, and caught it again. This was repeated, without a miss.

3. Each sea-lion balanced on the tip of its nose, first a ball like a baseball, then a large ball two feet in diameter.

4. Each sea-lion climbed a double ladder of eight steps, and went down on the other side, *balancing a large ball on the end of its nose, without a miss.*

5. The trainer handed a ball to the sea-lion nearest him, who balanced it on his nose, walked with it to his box and climbed up.

6. Then another sea-lion walked over to him, and waited expectantly until sea-lion No. 1 tossed the ball to No. 2, who caught it on his nose, walked over to his box, climbed up, and presently tossed it to No. 3.

7. A silk hat was balanced on its rim.

8. A seal carrying a balanced ball scrambled upon a cylindrical basket and rolled it across the arena, after which other seals repeated the performance.

9. In the last act a flaming torch was balanced, tossed about, caught and whirled, and finally returned to the trainer, still blazing.

Trained Horses. By carefully selecting the brightest and most intelligent horses that can be found, it is possible for a trainer to bring together and educate a group that will go through a fine performance in public. However, some exhibitions of trained horses are halting, ragged and poor. I have seen only one that stands out in my records as superlatively fine,—for horses. That was known to the public when I saw it as Bartholomew's "Equine Paradox," and it contained twelve wonderfully trained horses. My record of this fine performance fills seven pages of a good-sized notebook. While it is too long to reproduce here entire, it can at least be briefly described. The trainer called his group a "school," and of it he said:

"While I do not say that any one horse knows the meaning of from 300 to 400 words, I claim that *as a whole* the school does know that number."

The performance was fairly bewildering; but by diligent work I recorded the whole of it. Various horses did various things. They fetched chairs, papers, hats and coats; opened desks, rang bells, came when called, bowed, knelt, and erased figures from a blackboard. They danced a waltz, a clog dance, a figure-8; they marched, halted, paced, trotted, galloped, backed, jumped, leaped over each other, performed with a barrel, a see-saw and a double see-saw. Their marching and drilling would have been creditable to a platoon of rookies.

In performing, every horse is handicapped by his lack of hands and plant grade feet; and the horse memory is not very sure or certain. More than any other animal, the horse depends upon the trainer's command, and in poor performances the command often requires to be repeated, two or three times, or more. The memory of the horse is not nearly so quick nor so certain as that of the chimpanzee or elephant.

Dr. Martin J. Potter, of New York, famous trainer of stage and movie animals, states that of all animals, wild or domestic, the horse is the most intelligent; but I doubt whether he ever trained any chimpanzees. Speaking from out of the abundance of his training experience with many species of animals except the great apes, Dr. Potter says that "the seal [i. e. California sea-lion] learns its stage cues more easily than any other mute performer. The horse, however, is the most intelligent of all animals in its grasp and understanding of the work it has learned to perform, and in its reliable faithfulness and memory." Dr. Potter holds that of wild animals the tiger, owing to its treachery and ferocity, is the most difficult wild animal to train; the lion is the most reliable, and the most stupid of all animals is the pig.

The Taming of Boma. A keeper for a short time in our place, named D'Osta, once did a very neat piece of work in taming a savage and intractable chimpanzee. When Boma came to us, fresh from the French Congo, he was savage and afraid. He retreated to the highest resting-place of his cage, came down only at night for his meals, and would make no compromise. We believed that he had been fearfully abused by his former owners, and through mistreatment had acquired both fear and hatred of all men.

After the lapse of several months with Boma on that basis, the situation grew tiresome and intolerable. So D'Osta said:

"I must tame that animal, and teach him not to be afraid of us."

He introduced a roomy shifting cage into Boma's compartment, fixed the drop door, and for many days served Boma's food and water in that cage only. For two weeks the ape eluded capture, but eventually the keeper caught him. At first Boma's rage and fear were boundless; but presently the idea dawned upon his mind that he was not to be killed immediately. D'Osta handed him excellent food and water, twice a day, spoke to him soothingly, and otherwise let him alone. Slowly Boma's manner changed. He learned that he was not to be hurt, nor even annoyed. Confidence in the men about him began to come to him. His first signs of friendliness were promptly met and cultivated.

At the end of ten days, D'Osta opened the sliding door, and Boma walked out, a wiser and better ape. His bad temper and his fears were gone. He trusted his keeper, and cheerfully obeyed him. Strangest of all, he even suffered D'Osta to put a collar upon him, and chain him to the front bars to curb his too great playfulness while his cage was being cleaned.

Boma's fear of man has never returned. Now, although he is big and dangerous, he is a perfectly normal ape.

The Training of an Over-Age Bear. A bear-trainer-athlete and "bear-wrestler" named Jacob Glass once taught me a lesson that astounded me. It related to the training of a bear that I thought was too old to be trained.

We had an Alaskan cinnamon bear, three years old, that had been christened "Christian," at Skagway, because it stood so much pestering without flying into rages, as the grizzly did. After a short time with us, the concrete floors of our bear dens reacted upon the soles of its feet so strangely and so seriously that we were forced to transfer the animal to a temporary cage that had a wooden floor. While I was wondering what to do with that bear, along came Mr. Glass, anxious and unhappy.

"My wrestling bear has died on me," he said, "and I've got to get another. You have got one that I would like to buy from you. It's the one you call Christian."

Very kindly I said, "That is a mighty fine bear, as to temper; but now he is entirely too old to train, and you couldn't do anything with him. He would be a loss to you."

"I've looked him over, and I like his looks. I think I can train him all right. You let me have him, and I'll make a fine performer of him."

"I know that you never can do it; but you may try him, and send him back when you fail."

Thus ended the first lesson; and I was sure that in a month Mr. Glass would beg me to take back the untrainable animal.

About one year later Glass appeared again, jubilant. At once he broke forth into eulogies of Christian; but one chapter would not be large enough to contain them. He had trained that bear, with outrageous ease and celerity, and hadimmediately taken him upon the stage as a professional jiu-jitsu wrestler. And really, the act was admirable. As a wrestler, the bear seemed almost as intelligent as the man. He knew the "left-hand half-nelson" as well as Glass, and he knew the following words, perfectly: "Right, left, half-nelson, strangle, head up, nose under arm, and hammer-lock."

[Illustration with caption: THE WRESTLING BEAR "CHRISTIAN" AND HIS
PARTNER]

Glass declared that this bear was more intelligent than any lion, or any other trained animal ever seen by him. He was wise in many ways besides wrestling, — in his friendship with Glass, with other bears, with other men, and with a dog. *He obeyed all orders willingly,* even permitting Glass to take his food away when he was eating; but he would not stand being punished with a whip or a stick! In response to that he would bite. However, he generously permitted himself to be *held down and choked, as a punishment,* after which he would be very repentant, and would insist upon getting into his partner's lap, — to show his good will.

Glass was enthusiastically certain that Christian could reason independently from cause to effect. He declared that his alertness of mind was so pronounced it was very rarely necessary to show him a second time how to do a given thing.

Training an Adult Savage Monkey. Once we had a number of Japanese red-faced monkeys, and one of the surplus adult males had a temper as red as his face. Mr. Wormwood, an exhibitor of performing monkeys, wished to buy that animal; but I declined to sell it, on the ground that it would be impossible to train it.

At that implied challenge the trainer perked up and insisted upon having that particular bad animal; so we yielded. He wished him for the special business of turning somersaults, because he had no tail to interfere with that performance.

Two months later Mr. Wormwood appeared again. "Yes," he said, but not boastfully, "*I trained him;* but I came mighty near to giving him up as a bad job. He was the hardest subject I ever tackled; but I conquered him at last, and now he is working all right."

A really great number of different kinds of animals have been trained for stage performances, running the scale all the way up from fleas to elephants. It is easy to recall mice, rats, rabbits, squirrels, parrots, macaws, cockatoos, crows, chickens, geese, cats, pigs, dogs, monkeys, baboons, apes, bears, seals, sea-lions, walruses, kangaroos, horses, hippopotami and elephants. It is a large subject, and its many details are full of interest. It is impossible to discuss here all these species and breeds.

In concluding these notes I leave off as I began,—with the statement that any student of animal psychology who for any reason whatever ignores or undervalues the intelligence of trained animals puts a handicap upon himself.

III. THE HIGHER PASSIONS

XVIII

THE MORALS OF WILD ANIMALS

The ethics and morals of men and animals are thoroughly comparative, and it is only by direct comparisons that they can be analyzed and classified. It is quite possible that there are quite a number of intelligent men and women who are not yet aware of the fact that wild animals *have* moral codes, and that on an average they live up to them better than men do to theirs.

It is a painful operation to expose the grinning skeletons in the closets of the human family, but in no other way is it possible to hold a mirror up to nature. With all our brightness and all our talents,—real and imitation,—few men ever stop to ask what our horses, dogs and cats think of our follies and our wickedness.

By the end of the year 1921 the annual total of human wickedness had reached staggering proportions. From August 1914 to November 1918 the moral standing of the human race reached the lowest depth it ever sounded since the days of the cave-dwellers. This we know to be true, because of the increase in man's capacity for wickedness, and its crop of results. After what we recently have seen in Europe and Asia, and on the high seas, let no man speak of a monster in human form as "a brute;" for so far as moral standing is concerned, some of the animals allegedly "below man" now are in a position to look down upon him.

It is a cold and horrid fact that today, all around us, and sometimes close at hand, men are committing a long list of revolting crimes such as even the most debased and cruel beasts of the field *never* commit. I refer to wanton wholesale murder, often with torture; assault with violence, robbery in a hundred cruel forms, and a dozen unmentionable crimes invented by degenerate man and widely practiced. If anyone feels that this indictment is too strong, I can cite a few titles that will be quite sufficient for my case.

Let us make a few comparisons between the human species (*Homo sapiens*) and the so-called "lower" wild animals; and let it be understood that the author testifies, in courtroom phrase, only "to the best of his information and belief."

Only two wild animal species known to me, — wolves and crocodiles, — devour their own kind; but many of the races of men have been cannibals, and some are so today.

Among free wild animals, the cruel abuse or murder of children by their parents, or by other adults of the tribe, is unknown; but in all the "civilized" races of men infanticide and child murder are frightfully common crimes. In 1921 a six-year-old Eskimo girl, whose father and mother had been murdered, was strangled by her relatives, because she had no visible means of support.

The murder of the aged and helpless among wild animals is almost unknown; but among both the savage and the civilized races of men it is quite common. Our old acquaintance, Shack-Nasty Jim, the Modoc Indian, tomahawked his own mother because she hindered his progress; but many persons in and around New York have done worse than that.

Civil war between the members of a wild animal species is a thing unknown in the annals of wild-animal history; but among men it is an every-day occurrence.

Among *free* animals it is against the moral and ethical codes of all species of vertebrates for the strong to bully and oppress the weak; but it is almost everywhere a common rule of action with about ten per cent of the human race.

The members of a wild animal species are in honor bound not to rob one another, but with 25 per cent of the men of all civilized races, robbery, and the desire to get something for nothing, are ruling passions. No wild animals thus far known and described practice sex crimes; but the less said of the races of men on this subject, the better for our feelings.

Among animals, save in the warfare of carnivorous animals for their daily food, there are no exterminatory wars between species, and even local wars over territory are of very rare occurrence. Among men, the territorial wars of tribes and nations are innumerable, they have been from the earliest historic times, and they are certain to continue as long as this earth is inhabited by man. The "end of war" between the grasping nations of this earth is an iridescent dream, because of the inextinguishable jealousy and meanness of nations; but it is well to reduce them to a minimum. Nations like Germany, Bulgaria, Turkey and Russia will never stand hitched for any long periods. Their peace-loving neighbors need to keep their weapons well oiled and polished, and indulge in no hallucinations of a millenium upon this wicked earth.

In the mating season, there is fighting in many wild animal species between the largest and finest male individuals for the honor of overlordship in increasing and diffusing the species. These encounters are most noticeable in the various species of the deer family, because the fatal interlocking of antlers occasionally causes the death of both contestants. We have in our National Collection of Heads and Horns sets of interlocked antlers of moose, caribou, mule deer and white-tailed deer.

Otherwise than from the accidental interlocking of antlers, — due to the fact that an animal can push forward with far greater force than it can pull back, — I have never seen, heard or read of a wild animal having been *killed* outright in a fight over the possession of females. Fur seal and Stellar sea-lion bulls, and big male orang-utans, frequently are found badly scarified by wounds received in fighting during the breeding season, but of actual deaths we have not heard.

The first law of the jungle is: "Live, and let live."

Leaving out of account the carnivorous animals who must kill or die, *all the wild vertebrate species of the earth have learned the logic that peace promotes happiness, prosperity and long life.* This fundamentally useful knowledge governs not only the wild animal individual, but also the tribe, the species, and contiguous species.

Do the brown bears and grizzlies of Alaska wage war upon each other, species against species? By no means. It seems reasonably certain that those species occasionally intermarry. Do the big sea-lions and the walruses seek to drive away or exterminate the neighboring fur seals or the helpless hair seals? Such warfare is absolutely unknown. Do the moose and caribou of Alaska and Yukon Territory attack the mountain sheep and goats? Never. Does the Indian elephant attack the gaur, the sambar, the axis deer or the muntjac? The idea is preposterous. Does any species of giraffe, zebra, antelope or buffalo attack any other species on the same crowded plains of British East Africa? If so, we have yet to learn of it.

If the races and nations of men were as peace-loving, honest and sensible in avoiding wars as all the wild animal species are, then would we indeed have a social heaven upon earth.

Now, tell me, ye winged winds that blow from the four corners of the earth and over the seven seas, whence came the Philosophy of Peace to the world's wild animals? Did they learn it by observing the ways of man? "It is to laugh," says the innkeeper. Man has not yet learned it himself; and therefore do we find the beasts of the field a lap ahead of the quarrelsome biped who has assumed dominion over them.

Day by day we read in our newspapers of men and women who are moral lepers and utterly unfit to associate with horses, dogs, cats, deer and elephants. Our big male chimpanzee, Father Boma, who knows no wife but Suzette, and firmly repels the blandishments of his neighbor Fanny, is a more moral individual than many a pretty gentleman whose name we see heading columns of divorce proceedings in the newspapers.

Said the Count to Julia in "The Hunchback," "Dost thou like the picture, dearest?" As a natural historian, it is our task to hew to the line, and let the chips fall where they will.

Among the wild animals there are but few degenerate and unmoral species. In some very upright species there are occasionally individual lapses from virtue. A famous case in point is the rogue elephant, who goes from meanness to meanness until he becomes unbearable. Then he is driven out of the herd; he becomes an outcast and a bandit, and he upsets carts, maims bullocks, tears down huts and finally murders natives until the nearest local sahib gets after him, and ends his career with a bullet through his wicked brain.

In my opinion the gray wolf of North America (like his congener in the Old World) is the most degenerate and unmoral mammal species on earth. He murders his wounded packmates, he is a greedy cannibal, he will attack his wife and chew her unmercifully. On the other hand, his one redeeming trait is that he helps to rear the pups, — when they are successfully defended from him by their mother!

The wolverine makes a specialty of devilish and uncanny cunning and energy in destroying the property of man. Trappers have told us that when a wolverine invades a trapper's cabin in his absence, he destroys very nearly its entire contents. The food that he can neither eat nor carry away he defiles in such a manner that the hungriest man is unable to eat it. This seems to be a trait of this species only, — among wild animals; but during the recent war it was asserted that similar acts were committed by soldiers in the captured and occupied villas of northern France.

The domestication of the dog has developed a new type of animal criminal. The sheep-killing dog is in a class by himself. The wild dog hunts in the broad light of day, often running down game by the relay system. The sheep-killing dog is a cunning night assassin, a deceiver of his master, a shrewd hider of criminal evidence, a sanctimonious hypocrite by day but a bloody-minded murderer under cover of darkness. Sometimes his cunning is almost beyond belief. Now, can anyone tell us how much of this particular evolution is due to the influence of Man upon Dog through a hundred generations of captivity and association? Has the dog learned from man the science of moral banditry, the best methods for the concealment of evidence, and how to dissemble?

Elsewhere a chapter is devoted to the crimes of wild animals; but the great majority of the cases cited were found among *captive* animals, where abnormal conditions produced exceptional results. The crimes of captive animals are many, but the crimes of free wild animals are comparatively few. Whenever we disturb the delicate and precise balance of nature we may expect abnormal results.

XIX

THE LAWS OF THE FLOCKS AND THE HERDS

Through a thousand generations of breeding and living under natural conditions, and of self-maintenance against enemies and evil conditions, the wild flocks and herds of beasts and birds have evolved a short code of community laws that make for their own continued existence.

And they do more than that. When free from the evil influences of man, those flock-and-herd laws promote, and actually produce, peace, prosperity and happiness. This is no fantastic theory of the friends of animals. It is a fact, just as evident to the thinking mind as the presence of the sun at high noon.

The first wild birds and quadrupeds found themselves beset by climatic conditions of various degrees and kinds of rigor and destructive power. In the torrid zone it took the form of excessive rain and humidity, excessive heat, or excessive dryness and aridity. In the temperate and frigid zones, life was a seasonal battle with bitter cold, torrents of cold rain in early winter or spring, devastating sleet, and deep snow and ice that left no room for argument.

At the same time, the species that were not predatory found themselves surrounded by fangs and claws, and the never-ending hunger of their owners. The air, the earth and the waters swarmed with predatory animals, great and small, ever seeking for the herbivorous and traitorous species, and preferably those that were least able to fight or to flee. The La Brea fossil beds at Los Angeles, wherein a hospitable lake of warm asphalt conserved skeletal remains of vertebrates to an extent and perfection quite unparalleled, have revealed some very remarkable conditions. The enormous output, up to date, of skulls of huge lions, wolves, sabre-toothed tigers, bears and other predatory animals, shows, for once, just what the camels, llamas, deer, bison and mammoths of those days had to do, to be, and to suffer in order to survive.

With the aid of a little serious study, it is by no means difficult to recognize the hard laws that have enabled the elephant, bison, sheep, goats, deer, antelope, gazelles, fur-seal, walrus and others to survive and increase. From the wild animal herds and bird flocks that we have seen and personally known, *we know what their laws are,* and can set them down in the order of their evolution and importance.

The First Law. *There shall be no fighting in the family, the herd or the species, at any other time than in the mating season; and then only between adult males who fight for herd leadership.*

The destructiveness of intertribal warfare, either organized or desultory, must have been recognized in Jurassic times, millions of years ago, by the reptiles of that period. Throughout the animal kingdom below man the blessings of peace now are thoroughly known. This first law is obeyed by all species save man. We doubt whether all the testimony of the rocks added together can show that one wild species of vertebrate life ever really was exterminated by another species, not even excepting the predatory species which lived by killing.

No one (so far as we know) has charged that the lions, or the tigers, the bears, the orcas, the eagles or the owls have ever obliterated a species during historic times. It was the swine of civilization, transplanted by human agencies, that exterminated the dodo on the Island of Mauritius; and it was men, not birds of prey, who swept off the earth the great auk, the passenger pigeon and a dozen other bird species.

The Second Law. *The strong members of a flock or herd shall not bully nor oppress the weak.*

This law, constantly broken by degenerate and vicious men, women and children, very rarely is broken in a free wild herd or flock. In the observance of this fundamental law, born of ethics and expediency, mankind is far behind the wild animals. It would serve a good purpose if the criminologists and the alienists would figure out the approximate proportion of the human species now living that bullies and maltreats and oppresses the weak and the defenseless. At this moment "society" in the United States is in a state of thoroughly imbecilic defenselessness against the new type of predatory savages known as "bandits."

The Third Law. *During the annual period of motherhood, both prospective and actual, mothers must be held safe from all forms of molestation; and their young shall in no manner be interfered with.*

For the perpetuation of a family, a clan or a species, the protection of the mothers, and their weak and helpless offspring is a necessity recognized by even the dullest vertebrate animals. As birth-time or nesting-time approaches the wild flocks and herds universally permit the potential mothers to seek seclusion, and to work out their respective problems according to their own judgment and the means at their command. The coming mother looks for a spot that will afford (1) a secure hiding-place, (2) the best available shelter from inclement weather, (3) accessible food and water, and (4) cover or other protection for her young.

During this period the males often herd together, and they serve a protective function by attracting to themselves the attacks of their enemies. For the mothers, the bearing time is a truce time. There are fox-hunters who roundly assert that in spring fox hounds have been known to refuse to attack and kill foxes about to become mothers.

The Fourth Law. *In union there is strength; in separation there is weakness; and the solitary animal is in the greatest danger.*

It was the wild species of mammals and birds who learned and most diligently observed this law who became individually the most numerous. A hundred pairs of eyes, a hundred noses and a hundred pairs of listening ears increase about ten times the protection of the single individual against surprise attacks. The solitary elephant, bison, sheep or goat is far easier to stalk and approach than a herd, or a herd member. A wolf pack can attack and kill even the strongest solitary musk-ox, bison or caribou, but the horned herd is invincible. A lynx can pull down and kill a single mountain sheep ram, but even the mountain lion does not care to attack a herd of sheep. It is due solely to the beneficent results of this clear precept, and the law of defensive union, that any baboons are today alive in Africa.

The grizzly bear loves mountain-goat meat; but he does not love to have his inner tube punctured by the deadly little black skewers on the head of a billy. It is the Mountain Goats' Protective Union that condemns the silvertip grizzly to laborious digging for humble little ground-squirrels, instead of killing goats for a living. The rogue elephant who will not behave himself in the herd, and will not live up to the herd law, is expelled; and after that takes place his wicked race is very soon ended by a high- power bullet, about calibre .26. The last one brought to my notice was overtaken by Charles Theobald, State Shikaree of Mysore, in a Ford automobile; and the car outlived the elephant.

The Fifth Law. *Absolute obedience to herd leaders and parents is essential to the safety of the herd and of the individual; and this obedience must be prompt and thorough.*

Whenever the affairs of grown men and women are dominated by ignorant, inexperienced and rash juniors, look out for trouble; for as surely as the sun continues to shine, it will come. With an acquaintance that comprehends many species of wild quadrupeds and birds, I do not recall even one herd or flock that I have seen led by its young members. There are no young spendthrifts among the wild animals. For them, youthful folly is too expensive to be tolerated. The older members of the clan are responsible for its safety, and therefore do they demand obedience to their orders. They have their commands, and they have a sign language by which they convey them in terms that are silent but unmistakable. They order "Halt," and the herd stops, at once. At the command "Attention," each herd member "freezes" where he stands, and intently looks, listens and scents the air. At the order "Feed at will," the tension slowly relaxes; but if the order is "Fly!" the whole herd is off in a body, as if propelled by one mind and one power.

My first knowledge of this law of the flock came down to me from the blue ether when I first saw, in my boyhood, a V-shaped flock of Canada geese cleaving the sky with straight and steady flight, and perfect alignment. Even in my boyish mind I realized that the well-ordered progress of the wild geese was in obedience to Intelligence and Flock Law. Later on, I saw on the Jersey sands the mechanical sweeps and curves and doubles of flying flocks of sandpipers and sanderlings, as absolutely perfect in obedience to their leaders as the slats of a Venetian blind.

A herd of about thirty elephants, under the influence of a still alarm and sign signals, once vanished from the brush in front of me so quickly and so silently that it seemed uncanny. One single note of command from a gibbon troop leader is sufficient to set the whole company in instant motion, fleeing at speed and in good order, with not a sound save the swish of the small branches that serve as the rungs of their ladder of flight.

In the actual practice of herd leadership in species of ruminant animals, the largest and most spectacular bull elk or bison is not always the leader. Frequently it has been observed that a wise old cow is the actual leader and director of the herd, and that "what she says, goes." This was particularly remarked to me by James McNaney during the course of our "last buffalo hunt" in Montana, in 1886. From 1880 to 1884 he had been a mighty buffalo-hunter, for hides. He stated that whenever as a still-hunter he got "a stand on a bunch," and began to shoot, slowly and patiently, so as not to alarm the stand, whenever a buffalo took alarm and attempted to lead away the bunch, usually it proved to be a wise old cow. The bulls seemed too careless to take notice of the firing and try to lead away from it.

The Sixth Law. *Of food and territory, the weak shall have their share.*

While this law is binding upon all the members of a wild flock, a herd, a clan or a species, outside of species limits it may become null and void; though in actual practice I think that this rarely occurs. Among the hoofed animals; the seals and sea-lions; the apes, baboons and monkeys, and the kangaroos, the food that is available to a herd is common to all its members. We can not recall an instance of a species attempting to dispossess and evict another species, though it must be that many such have occurred. In the game-laden plains of eastern Africa, half a dozen species, such as kongonis, sable antelopes, gazelles and zebras, often have been observed in one landscape, with no fighting visible.

With all but the predatory wild animals and man, the prevailing disposition is to *live, and let live.* One of the few recorded murders of young animals by an old one of the same species concerned the wanton killing of two polar bear cubs in northern Franz Joseph Land, as observed by Nansen.

The Seventh Law. *Man is the deadliest enemy of all the wild creatures; and the instant a man appears the whole herd must fly from him, fast and far.*

In some of the regions to which man and his death-dealing influence have not penetrated, this law is not yet on the statute books of the jungle and the wilderness. Sir Ernest Shackleton and Captain Scott found it unknown to the giant penguins and sea leopards of the Antarctic Continent, I have seen a few flocks and herds by whom the law was either unknown or forgotten; but the total number is a small one. There was a herd of mountain sheep on Pinacate Peak, a big flock of sage grouse in Montana, various flocks of ptarmigan on the summits of the Elk River Mountains, British Columbia,—and out of a long list of occurrences that is all I will now recall.

It is fairly common for the members of a vast assemblage of animals, like the bison, barren-ground caribou, fur seal, and sea birds on their nesting cliffs, to assume such security from their numbers as to ignore man; and all such cases are highly interesting manifestations of the influence of the fourth law when carried out to six decimal places.

The Eighth and Last Law. *Whenever in a given spot all men cease to kill us, there may we accept sanctuary and dwell in peace.*

This law comes as Amendment 1 to the original Constitution of the Animal Kingdom. The quick intelligence of wild animals in recognizing a new sanctuary, and in adopting it unreservedly and thankfully as their own territory, is to all friends of wild life a source of wonder and delight. With their own eyes Americans have seen the effects of sanctuary-making upon bison, elk, mule deer, white-tailed deer, mountain sheep, mountain goat, prong-horned antelope, grizzly and black bears, beavers, squirrels, chipmunks, rabbits, sage grouse, quail, wild ducks and geese, swans, pelicans brown and white, and literally hundreds of species of smaller birds of half a dozen orders.

In view of this magnificent and continent-wide manifestation of discovery, new thought and original conclusion, let no man tell us that the wild birds and quadrupeds "do not think" and "can not reason."

The Exceptions of Captivity. When wild animals come into captivity, a few individuals develop and reveal their worst traits of character, and much latent wickedness comes to the surface. A small percentage of individuals become mean and lawless, and a still smaller number show criminal instincts. These Bolshevistic individuals commit misdemeanors and crimes such as are unknown in the wild state. One male ruminant out of perhaps fifty will turn murderer, and kill a female or a fawn, entirely contrary to the herd law; and at long intervals a male predatory animal kills his mate or young.

Occasionally captivity warps wild animal or wild bird character quite out of shape, though it is a satisfaction to know that the total proportion of those so affected is very small. Long and close confinement in a prison-like home, filled with more daily cares and worries than any animal cage has of iron bars, has sent many a human wife and mother to an insane asylum; but the super- humanitarians who rail out at the existence of zoological parks and zoos are troubled by that not at all.

XX

PLAYS AND PASTIMES OF ANIMALS

I approach this subject with a feeling of satisfaction; but I would not like to state the number of hours that I have spent in watching the play of our wild animals.

Out in the wilds, where the bears, sheep and goats live and thrive, the outdoorsmen see comparatively few wild animals at play. No matter what the season, the dangers of the wilderness and mountain summit remain the same. When kids and lambs are young, the eaglets are hungriest, and their mothers are most determined in their hunting. After September 1, the deadly still-hunters are out, and strained watchfulness is the unvarying rule, from dawn until dark.

Out in the wilds, it is the *moving* animal that instantly catches every hostile eye within visual range. A white goat kid vigorously gamboling on the bare rocks would attract all the golden eagles, hunters, trappers and Indians within a radius of two miles. It is the rule that kids, fawns and lambs must *lie low and keep still,* to avoid attracting deadly enemies. On the bare summits, play can be indulged in only at great risk. Generations of persecution have implanted in the brain of the ruminant baby the commanding instinct to fold up its long legs, neatly and compactly, furl its ears along its neck, and closely lie for hours against a rock or a log. During daylight hours they must literally hug the ground. Silence and inactivity is the first price that all young animals in the wilds pay for their lives. It is only in the safe shelter of captivity, or man-made sanctuaries, that they are free to play.

In the comfortable security of the "zoo" all the wild conditions are changed. The restraints of fear are off, and every animal is free to act as joyous as it feels. Here we see things that men *never see in the wilds!* If any Rocky Mountain bear hunter should ever see bear cubs or full-grown bears wrestling and carrying on as they do here, he would say that they were plumb crazy!

Of all our wild animals, not even excepting the apes and monkeys, our young bears are the most persistently playful. In fact, I believe that when *properly caged and tended,* bears under eight years of age are the most joyous and playful of all wild animals. We have given our bears smooth and spacious yards floored with concrete, with a deep pool in the centre of each, and great possibilities in climbing upon rocks high and low. The top of each sleeping den is a spacious balcony with a smooth floor. The facilities for bear wrestling and skylarking are perfect, and there are no offensive uneven floors nor dead stone walls to annoy or discourage any bear. They can look at each other through the entire series of cages and there is no chance whatever for a bear to feel lonesome. We put just as many individuals into each cage as we think the traffic will stand; and sometimes as many as six young bears are reared together.

Now, all these conditions promote good spirits, playfulness, and the general enjoyment of life. Any one who thinks that our bears are not far happier than those that are in the wilds and exposed to enemies, hunger and cold, should pause and consider.

Our bear cubs begin to play just as soon as they emerge from their natal den, in March or April, and they keep it up until they are six or seven years of age, — or longer! Our visitors take the playfulness of small cubs as a matter of course, but the clumsy and ridiculous postures and antics of fat-paunched full-grown bears are irresistibly funny. Really, there are times when it seems as if the roars of laughter from the watching crowd stimulates wrestling bears to further efforts. On October 28, 1921, about seventy boys stood in front of and alongside the den of two Kluane grizzly cubs and shouted for nearly half an hour in approval and admiration of the rapid and rough play of those cubs.

[Illustration with caption: ADULT BEARS AT PLAY]

The play of bears, young or middle-aged, consists in boxing, catch-as-catch-can wrestling, and chasing each other to and fro. Cubs begin to spar as soon as they are old enough to stand erect on their hind feet. They take their distance as naturally as prize-fighters, and they strike, parry and dodge just as men do. They handle their front feet with far more dexterity and precision than boys six years of age.

Boxing bears always strike for the head, and bite to seize the cheek of the opponent. In biting, mouth meets mouth, in defense as well as attack. When a biting bear makes a successful pass and finally succeeds in getting a firm toothhold on the cheek of his opponent, the party of the second part promptly throws himself prone upon the ground, and with four free feet concentrated upon the head of the other bear forces him to let go. This movement, and the four big, flat foot soles coming up into action is, in large bears, a very laughable spectacle, and generally produces a roar.

Wrestling bears roll over and over on the ground, clawing and biting, until one scrambles up, and either makes a new attack or rushes away.

Bears love to chase one another, *and be chased;* and in this form of skylarking they raise a whirlwind of activity which leads all around the floor, up to the balcony and along the length of it, and plunges down at the other end. Often a bear that is chased will fling himself into the bathing pool, with a tremendous splash, quickly scramble out again and rush off anew in a swirl of flying water.

The two big male polar bears that came to us from the William Hagenbeck group were very fond of playing and wrestling in the water of their swimming pool. Often they kept up that aquatic skylarking for two hours at a stretch, and by this constant claw work upon each other's pelts they kept their coats of hair so thinned down that we had to explain them. One bear had a very spectacular swimming trick. He would swim across the pool until his front feet touched the side, then he would throw himself over backwards, put his hind feet against the rock wall, and with a final shove send himself floating gracefully on his back across to the other side.

Playful bears are much given to playing tricks, and teasing each other. A bear sleeping out in the open den is regarded as a proper subject for hectoring, by a sudden bite or cuff, or a general assault. It is natural to expect that wrestling bears will frequently become angry and fight; but such is not the case. This often happens with boys and men, but bears play the game consistently to the end. I can not recall a single instance of a real bear fight as the result of a wrestling or boxing match; and may all boys take note of this good example from the bear dens.

Next to the bears, the apes and monkeys are our most playful animals. Here, also, it is the young and the half grown members of the company that are most active in play. Fully mature animals are too sedate, or too heavy, for the frivolities of youth. A well- matched pair of young chimpanzees will wrestle and play longer and harder than the young of any other primate species known to me. It is important to cage together only young apes of equal size and strength, for if there is any marked disparity in size, the larger and stronger animal will wear out the strength of its smaller cage-mate, and impair its health.

In playing, young chimps, orangs or monkeys seize each other and wrestle, fall, and roll over and over, indefinitely. They make great pretenses of biting each other, but it is all make-believe. My favorite orang-utan pet in Borneo loved to play at biting me, but whenever the pressure became too strong I would say chidingly, "Ah! Ah!" and his jaws would instantly relax. He loved to butt me in the chest with his head, make wry faces, and make funny noises with his lips. I tried to teach him "cat's cradle" but it was too much for him. His clumsy fingers could not manage it.

One of our brightest chimpanzees, named Baldy, was much given to hectoring his female cage-mate, for sport. What he regarded as his best joke was destroying her bed. Many times over, after she had laboriously carried straw up to the balcony, carefully made up a nice, soft, circular bed for herself, and settled down upon it for a well-earned rest, Baldy would silently climb up to her level, suddenly fling himself upon her as she lay, and with all four of his arms and legs violently working, the nest would be torn to pieces and scattered and the lady orang rudely pulled about. Then Baldy would joyously swing down to the lower level, settle himself demurely at the front of the cage, and with a placid face and innocent, far-away expression in his eyes gaze at the crowd. There was nothing lacking but a mischievous wink of one eye.

Whenever his cage-mate selected a particularly long and perfect straw and placed it crosswise in her mouth, Baldy would steal up behind her and gleefully snatch it away.

Baldy was a born comedian. He loved to amuse a crowd and make people laugh. He would go through a great trapeze performance of clownish and absurd gymnastics, and often end it with three or four loud smacks of his big black feet against the wall. This was accomplished by violent kicking backwards. His dancing and up-and-down jumping always made visitors laugh, after which he would joyously give his piercing *"Wah-hoo"* shout of triumph. A Sioux Indian squaw dances by jumping up and down, but her performance is lifeless in comparison.

No vaudeville burlesque dancer ever cut a funnier monkey shine than the up-and-down high-jump dance and floor-slapping act of our Boma chimpanzee (1921). Boma offers this whenever he becomes especially desirous of entertaining a party of distinguished visitors. In stiff dancing posture, he leaps high in the air, precisely like a great black jumping-jack straight from Dante's Inferno. Orangs love to turn somersaults, and some individuals are so persistent about it as to wear the hair off their backs, disfigure their beauty, and disgust their keepers.

In the chapter on "Mental Traits of the Gorilla" a descriptionis given of the play of Major Penny's wonderful John Gorilla.

When many captive monkeys are kept together in one large cage containing gymnastic properties, many species develop humor, and indulge in play of many kinds. They remind me of a group of well-fed and boisterous small boys who must skylark or "bust." From morning until night they pull each other's tails, wrestle and roll, steal each other's playthings, and wildly chase each other to and fro. There is no end of chattering, and screeching, and funny facial grimaces. A writer in *Life* once said that the sexes of monkeys can be distinguished by the fact that "the females chatter twice as fast as the males," but I am sure that many ladies will dispute that statement.

In a company of mixed monkeys, or a mixed company of monkeys, a timid and fearsome individual is often made the butt of practical jokes by other monkeys who recognize its weakness. And who has not seen the same trait revealed in crowds of boys?

But we can linger no longer with the Primates.

Who has not seen squirrels at play? Once seen, such an incident is not soon forgotten. I have seen gray, fox and red squirrels engage in highly interesting performances. The gray squirrel is stately and beautiful in its play, but the red squirrel is amazing in its elaborateness of method. I have seen a pair of those mischief- makers perform low down on the trunk of a huge old virgin white oak tree, where the holding was good, and work out a program almost beyond belief. They raced and chased to and fro, up, down and across, in circles, triangles, parabolas and rectangles, until it was fairly bewildering. Really, they seemed to move just as freely and certainly on the tree-trunk as if they were on the ground, with no such thing in sight as the law of gravitation.

It seems to me that the gray squirrel barks and the red squirrel chatters, scolds, and at times swears, chiefly for the fun of hearing himself make a noise. In the red squirrel it is impudent and defiant; and usually you hear it near your camp, or in your own grounds, where the rascals know that they will not be shot.

The playful spirit seems to be inherent in the young of all the Felidae. The playfulness of lion, tiger, leopard and puma cubs is irresistibly pleasing; and it is worth while to rear domestic kittens in order to watch their playful antics.

I have been assured by men who seemed to know, that wolf and fox cubs silently play in front of their home dens, when well screened from view, just as domestic dog puppies do; and what on earth can beat the playfulness of puppies of the right kind, whose parents have given them red blood instead of fat as their inheritance. Interesting books might be written about the play of dogs alone.

The play of the otter, in sliding down a long and steep toboggan slide of wet and slippery earth to a water plunge at the bottom, is well known to trappers, hunters, and a few naturalists. It is quite celebrated, and is on record in many places. I have seen otter slides, but never had the good luck to see one in use. The otters indulge in this very genuine sport with just as much interest and zest as boys develop in coasting over ice and snow with their sleds.

Here at the Zoological Park, young animals of a number of species amuse themselves in the few ways that are open to them. It is a common thing for fawns and calves of various kinds to butt their mothers, just for fun. A more common form of infantile ruminant sport is racing and jumping. Now and then we see a red buffalo calf three or four months old suddenly begin a spell of running for amusement, in the pure exuberance of health and good living. A calf will choose a long open course, usually up and down a gentle slope, and for two hundred feet or more race madly to and fro for a dozen laps, with tail stiffly and very absurdly held aloft. Of course men and beasts all pause to look at such performances, and at the finish the panting and perspiring calf halts and gazes about with a conscious air of pride. All this is deliberate "showing off," just such as small boys frequently engage in.

Elk fawns, and more rarely deer fawns, also occasionally indulge in similar performances. Often an adult female deer develops the same trait. One of our female Eld's deer annually engages in a series of spring runs. We have seen her race the full length of her corral, up and down, over a two hundred foot course, at really break-neck speed, and keep it up until her tongue hung out.

Years ago, in the golden days, I was so lucky as to see several times wonderful dances of flocks of saras cranes on the low sandy islets in the River Jumna, northern India, just below Etawah. It was like this: While the birds are idly stepping about, apropos of nothing at all, one suddenly flaps his long wings several times in succession, another jumps straight up in the air for a yard or so, and presto! with one accord the whole flock is galvanized into action. They throw aside their dignity, and real fun begins. Some stand still, heads high up, and flap their wings many times. Others leap in the air, straight up and down, one jump after another, as high as they can go. Others run about bobbing and bowing, and elaborately courtesying to each other with half opened wings, breasts low down and their tails high in the air, cutting very ridiculous figures.

In springtime in the Zoological Park we often see similar exhibitions of crane play in our large crane paddock. A particularly joyous bird takes a fit of running with spread wings, to and fro, many times over, and usually one bird thus performing inspires another, probably of his own kind, to join in the game. The other cranes look on admiringly and sometimes a spectator shrilly trumpets his approval.

In his new book, "The Friendly Arctic," Mr. Vilhjalmur Stefansson records an interesting example of play indulged in jointly by a frivolous arctic fox and eight yearling barren-ground caribou. It was a game of tag, or its wild equivalent. The fox ran into and through the group of caribou fawns, which gave chase and tried to catch the fox, but in vain. At last the fawns gave up the chase, returned to their original position, and came to parade rest. Then back came the fox. Again it scurried through the group in a most tantalizing manner, which soon provoked the fawns to chase the fox anew. At the end of this inning the caribou again abandoned the chase, whereupon the fox went off to attend to other affairs.

On the whole, the play of wild animals is a large field and no writer will exhaust it with one chapter. Very sincerely do we wish that at least one of the many romance writers who are so industriously inventing wild-animal blood-and-thunder stories would do more work with his eyes and less with his imagination.

XXI

COURAGE IN WILD ANIMALS

Either in wild animals or tame men, courage is the moral impulse that impels an individual to fight or to venture at the risk of bodily harm. Like Theodore Roosevelt, the truly courageous individual engages his adversary without stopping to consider the possible consequences to himself. The timid man shrinks from the onset while he takes counsel of his fears, and reflects that "It may injure me in my business," or that "It may hurt my standing;" and in the end he becomes a slacker.

Among the mental traits and passions of wild creatures, a quantitative and qualitative analysis of courage becomes a highly interesting study. We can easily fall into the error of considering that fighting is the all-in-all measure of courage; which very often is far from being true. The mother quail that pretends to be wounded and feigns helplessness in order to draw hostile attention unto herself and away from her young, thereby displays courage of a high order. No quail unburdened by a helpless brood requiring her protection ever dreams of taking such risks. The gray gibbons of Borneo, who quite successfully made their escape from us, but promptly returned close up to my party in response to the S. O. S. cries of a captured baby gibbon, displayed the sublime courage of parental affection, and of desperation. Wary, timid and fearfully afraid of man, at the first sight of a biped they swing away. At the first roar of a gun they literally fly down hill through the treetops, and vanish in a wild panic. And yet, the leading members of that troop halted and swiftly came back, piercing the gloom and silence of the forest with their shrill cries of mingled encouragement and protest. It was quite as courageous and heroic as the act of a father who rushes into a burning building to save his child, at the imminent risk of his own life.

The animal world has its full share of heroes. Also, it has its complement of pugilists and bullies, its cowards and its assassins.

Few indeed are the wild creatures that fight gratuitously, or attack other animals without cause. If a fight occurs, look for the motive. The wild creatures know that peace promotes happiness and long life. Now, of all wild quadrupeds, it is probable that the African baboons are pound for pound the most pugnacious, and the quickest on the draw. The old male baboon in his prime will fight anything that threatens his troop, literally at the drop of a hat. But there is method in his madness. He and his wives and children dwell on the ground in lands literally reeking with fangs and claws. He has to confront the lion, leopard, wild dog and hyena, and make good his right to live. No wonder, then, that his temper is hot, his voice raucous and blood-curdling; his canines fearfully long and sharp, and his savage yell of warning sufficient to keep even the king of beasts off his grass.

Once I saw two baboons fight. We had two huge and splendid adult male gelada baboons, from Abyssinia. They were kept separate, but in adjoining cages; and the time came when we needed one of those cages for another distinguished arrival. We decided to try the rather hazardous experiment of herding those two geladas together.

Accordingly, we first opened the doors to both outside cages, to afford for the moment a free circulation of baboons, and then we opened the partition door. Instantly the two animals rushed together in raging combat. With a fierce grip each seized the other by the left cheek; and then began a baboon cyclone. They spun around on their axis, they rolled over and over on the floor, and they waltzed in speechless rage over every foot of those two cages. Strange to say, beyond coughing and gasping they made no sounds. Never before had we witnessed such a fearsome exhibition of insane hatred and rage.

As soon as the horrified spectators could bring it about, the wild fighters were separated; and strange to say, neither of them was seriously injured. It was a drawn battle.

It is quite difficult to weigh and measure the independent and abstract courage inherent in any wild animal species. All that can be done is to grope after the truth. On this subject there can be almost as many different opinions as there are species of wild animals.

What animal will go farthest in daring and defying man, even the man with a gun, in foraging for food?

Unquestionably and indisputably, the lion. This is no idle repetition of an old belief, or tradition. It is a fact; and we say this quite mindful of the records made by the grizzly bear, the Alaskan brown bear, the tiger, the leopard and the jaguar.

"The Man-Eaters of Tsavo" opened up a strange and new chapter in the life history of the savage lion. That truthful record of an astounding series of events showed the lion in an attitude of permanent aggression, backed by amazing and persistent courage. For several months in that rude construction camp on the arid bank of the Tsavo River, where a railway bridge was being constructed on the famous Uganda Railway line of British East Africa, lions and men struggled mightily and fought with each other, with living men as the stakes of victory. The book written by Col. J.H. Patterson, under the title mentioned above, tells a plain and simple story of the nightly onslaughts of the lions, the tragedies suffered from them, the constant, the desperate though often ill-considered efforts of the white engineers to protect the terrorized black laborers, and finally the death of the man-eaters. During a series of battles lasting four long months the two lions *killed and carried of a total of twenty-eight men!* How many natives were killed and not reported never will be known. The most hair-raising episode of all had a comedy touch, and fortunately it did not quite end in a tragedy. This is what happened:

Col. Patterson and his staff decided to try to catch the boldest of the lions in a trap baited with *a living man.* Accordingly a two-room trap was built, one room to hold and protect the man-bait, the other to catch and hold the lion. A very courageous native consented to be "it," and he was put in place and fastened up. The lion came on schedule time, he found the live bait, boldly entered the trap to seize it, and the dropping door fell as advertised. When the lion found himself caught, did his capture trouble him? Not in the least. Instead of starting in to tear his way out he decided to postpone his escape until he had torn down the partition and eaten the man! So at the partition he went, with teeth and claws.

In mortal terror the live bait yelled for succor. In "the last analysis" the man was saved from the lion, but the lion joyously tore his way out and escaped without a scratch. So far from being daunted by this divertisement he continued his man-killing industry, quite as usual.

Now, the salient points of the man-eaters of Tsavo consist of the unquenchable courage of the two lions, and their persistent defiance of white men armed with rifles. I am sure that there is nowhere in existence another record of wild-animal courage equal to this, and the truthfulness of it is quite beyond question.

The annals of African travel and exploration contain instances innumerable of the unparalleled courage of the lion in taking what he wants when he wants it.

THE GRIZZLY BEAR'S COURAGE. As a subject, this is a hazardous risk, because so many men are able to tell all about it. Judging from reliable records of the ways and means of the grizzly bear, I think we must award the second prize for courage to "Old Ephraim." The list of his exploits in scaring pioneers, in attacking hunters, in robbing camps, and finally in bear- handling and almost killing two guides in the Yellowstone Park, is long and thrilling. The record reaches back to the days of Lewis and Clark, who related many wild adventures with bears. The grizzlies of their day were very courageous, but even then they were *not* greatly given to attacking men quite unprovoked! In those days of bow-and-arrow Indians, and of white men armed only with ineffective muzzle-loading pea rifles, using only weak black powder, the grizzlies had an even chance with their human adversaries, and sometimes they took first money. In those days the courage of the grizzly was at its highest peak; and it was then conceded by all frontiersmen that the grizzly was thoroughly courageous, and always ready to fight. In the light of subsequent history, and in order to be just to the grizzly, we claim that his fighting was *in self defense,* for even in those days the unwounded bear preferred to run rather than to fight unnecessarily.

The rise of the high-power, long-range repeating rifle has made the grizzly bear a different animal from what he was in the days of Lewis and Clark. He has learned, *thoroughly,* the supreme deadliness of man's new weapons, and he knows that he is no longer able to meet men on even terms. Consequently, he runs, he hides, he avoids man, everywhere save in the Yellowstone Park, where he has found out that firearms are prohibited. There he has broken the truce so often that his offenses have had to be met with stern disciplinary measures that have made for the safety of tourists and guides.

Once I saw an amusing small incident. Be it known that when a new black bear cub is introduced to a den of its peers, the newcomer shrinks in fright, and cowers, and takes its place right humbly. But species alter cases. Once when we received an eight-months- old grizzly cub we turned it loose in a big den that contained five black bear cubs a year older than itself. But did the grizzly cub cower and shrink? By no manner of means. With head fully erect, it marched calmly to the centre of the den, and with serene confidence gave the other cubs the once-over with an air that plainly said: "*I'm* a grizzly! I'm here, and I've come to stay. Do I hear any objections?"

Quite as if in answer to the challenge, an eighteen-months-old black bear presently sidled up and made a trial blow at the grizzly's head. Instantly the grizzly cub's right arm shot out a well-delivered blow that sent the black one scurrying away in a panic, and perceptibly cleared the atmosphere. That cub had grizzly-bear *courage* and *confidence;* that was all.

There are a number of American sportsmen who esteem the Cape buffalo as the most aggressive and dangerous wild animal in eastern Africa. He is so courageous and so persistently bold that he is much given to lying in wait for hunters and attacking with real fury. The high grass of his swamps is very helpful to him as a means of defense. In our National Collection of Heads and Horns there is a huge buffalo head (for years the world's highest record) that tells the story of a near tragedy. The brother of Mr. F.H. Barber, of South Africa, fired at the animal, but failed to stop it. His gun jammed, and the charging beast was almost in the act of killing him when F.H. Barber fired without pausing to take aim. His lucky bullet knocked a piece out of the buffalo's left horn, dazed the animal for a moment, and afforded time for the shot that killed the mighty bull.

The leopard is usually a vicious beast. When brought to bay it fights with great fury and success. The black leopard is supremely vicious and intractable. Nearly all leopards hate training, and I have seen two or three leopard "acts" that were nerve-racking to witness because of the clear determination of all the animals to kill their trainer at the first opportunity.

The status of the big Alaskan brown bear has already been referred to in terms that may stand as an estimate of its courage. Really, it is now in the same mental state as the grizzly bears of the days of Lewis and Clark, and the surplus must be shot to admonish the survivors and protect the rights of man.

THE RAGE OF A WILD BULL ELK. One of the most remarkable cases of rage, resentment and fighting courage in a newly captured wild animal occurred near Buttonwillow, California, in November 1904, and is very graphically described by Dr. C. Hart Merriam in the *Scientific Monthly* for November 1921. The story concerns the leader of a band of the small California Valley Elk (*Cervus nannodes*) which it was desired to transport to Sequoia Park, for permanent preservation.

The bull refused to be driven to the corral for capture, so he was roped, thrown, hog-tied and hauled six miles on a wagon. This indignity greatly enraged the animal. At the corral he was liberated for the purpose of driving him through a chute and into a car.

From his capture and the jolting ride the bull was furious, and he refused to be driven. His first act was to gore and mortally wound a young elk that unfortunately found itself in the corral with him. Then he was roped again and his horns were sawn off. At first no horseman dared to ride into the corral to attempt to drive the animal. Finally the leader of the cowboys, Bill Woodruff, mounted on a wise and powerful horse who knew the game quite as well as his rider, rode into the corral with the raging elk, and attempted to drive it.

The story of the fight that followed, of raging elk vs. horse and man, makes stories of Spanish bullfights seem tame and commonplace, and the adventure of St. George and the dragon a dull affair. With the stubs of his antlers the bull charged the horse again and again, inflicting upon the splendid animal heart-rending punishment. Finally, after a fearful conflict, the wise and brave horse conquered, and the elk devil was forced into the car.

After a short railway journey the elk was forced into a crate,— fighting at every step,—and hauled a two days' journey to the Park. Reduced to kicking as its sole expression of resentment, the animal kicked continuously for forty-eight hours, almost demolishing the crate.

The final scene of this unparalleled drama of wild-animal rage is thus described by Dr. Merriam: "Then the other gates were raised, giving the bull an opportunity to step out. For the first, time since his capture he did what was wanted; he voluntarily crept to the rear of the wagon and hobbled out on the ground. Looking around for an enemy to attack and not seeing any, —some of the men having stationed themselves outside the park fence, the others on top of the crate,—he set out for the river, only a few rods away.

"His courage had not forsaken him, but his strength had. He was no longer the proudly aggressive wild beast he had been. He had reached his limit. The terrible ordeal he had been through; the struggle incident to his capture; the rough, hot ride to the corral, hog-tied, on the hard floor of the dead-ax wagon; the outbursts of passion in the corral; the fighting and second roping in connection with the sawing off of his horns; the battle with the big horse; the ceaseless violence of his destructive assaults, first in the car, then in the crate, continued for three days and nights, had finally undermined even his iron frame; so when at last he found himself free on the ground, he presented a truly pitiful picture.

"With his head bent to one side and back curved, with one ear up and the other down, and with a dejected, helpless expression on his face, he hobbled wearily away, barely able to step without falling. Slowly he made his way to the river, waded in, drank, crossed to the far side, staggered laboriously up the low bank, and lay down. The next day he was found in the same spot, — dead."

THE DEFENSE OF THE HOME AND FAMILY. Any man who is too cowardly to fight for his home and country deserves to live and die homeless and without a country.

With this subject of courage the parental and fraternal affections of wild animals are inseparably linked. The defense of the home and family unit is the foundation of all courage, and of all fighting qualities in man or animals. The gospel of self-defense is the first plank in the platform of the home defenders. Obviously, the head of a family cannot permit himself to be knocked out, because as the chief fighter in the Home Defense League it is his bounden duty to preserve his strength and his weapons, and remain fit.

In the days of the club, the stone axe and the flint arrow-head, men were few and feeble, and the wild beasts had no cause to fear extermination. Tooth, claw and horn were about as formidable as the clumsy and inadequate weapons of man. The wild species went on developing naturally, and some mighty hosts were the result.

But gunpowder changed all that. In the chase it gave weak men their innings beside the strong. Man could kill at long range, with little danger to himself, or even with none at all. And then in the wild beast world the great final struggle for existence began. Man's flippant phrase, — "the survival of the fittest," — became charged with sinister and deadly meaning.

But for Mother Love among wild creatures, species would not multiply, and the earth soon would become depopulated. In the entire Deer Family of the world, the annual shedding of all horns is Nature's tribute to motherhood in the herd. A buck deer or a bull moose is a domineering master — so long as his antlers remain upon his head. But with the approach of fawn-bearing time in the herd, down they go. I have seen a bull elk stand with humbly lowered head, and gaze reproachfully upon his fallen antlers. The dehorned buck not only no longer hectors and drives the females, but in fear of hurting his tender new velvet stubs he keeps well away from the front hoofs of the cows. The calves grow up quite safe from molestation within the herd.

It may be set down as a basic truth that all vertebrate animals are ready to defend their homes and their young against all enemies that do not utterly outclass them in size and strength. Of course we do not expect the pygmy to try conclusions with the giant, but at the same time, wild creatures have their own queer ways of defense and counter-attack, and of matching superior cunning against superior force. But now, throughout the animal world, the fear of man is paramount. Nearly all the wild ones have learned it. It is only the enraged, the frightened or the cornered bear, lion, tiger or elephant that charges the Man with a Gun, and seeks to counter upon him with fang and claw before it drops. The deadly supremacy of the repeating rifle that kills big game at half a mile, and the pump shotgun that gets five geese out of a flock, are well recognized by the terrorized big game and small game that flies before the sweeping pestilence of machine guns and automobiles.

THE FIGHTING CANADA GOOSE. In essaying to illustrate the home defense spirit, my memory goes out to one truculent and fearless Canada goose whose mate elected to nest in a horribly exposed spot on the east bank of our Wild-Fowl Pond. The location was an error in judgment. As soon as the nest was finished and the eggs laid therein, the goose took her place upon the collection, and the gander mounted guard.

There were so many hostiles on the warpath that he was kept on the qui vive during all daylight hours. At a radius of about twenty feet he drew an imaginary dead-line around the family nest, and no bird, beast or man could pass that line without a fight. If any other goose, or a swan or duck, attempted to pass, the guardian gander would rush forward with blazing eyes, open beak, wings open for action, and with distended neck hiss out his challenge. If the intruder failed to register respect, and came on, the gander would seize the offender with his beak, and furiously wing-beat him into flight. That gander was afraid of nothing, and his courage and readiness to fight all comers, all day long, caused visitors to accord him full recognition as a belligerent power.

THE CASE OF THE LAUGHING GULL. About that same time, a pair of laughing gulls had the temerity to build a nest on the ground in the very storm centre of the great Flying Cage. Daily and hourly they were surrounded by a truculent mob of pelicans, herons, ibises, storks, egrets and ducks, the most of whom delighted in wrecking households. The keepers sided with the gulls by throwing around their nest a wire entanglement, with a sally-port at one side for the use of the beleaguered pair.

The voice of an angry or frightened laughing gull is it [sic] owner's chief defense. The female sat on her nest and shrieked out her shrill and defiant war cry of "Kah! kah, kah, kah!" The male took post just outside the sally-port, where he postured and screamed and threatened until we wondered why he did not burst with superheated emotion. I am sure that never before did two small gulls ever raise so much racket in so short a time and their cage-mates must have found it rather trying.

The gulls hatched their eggs, they reared their young successfully, and at last peace was restored.

A Mother Antelope Fights Off an Eagle. Mr. Howard Eaton, of Wolf, Wyoming, once saw a female prong-horned antelope put up a strong and successful fight in defense of her newly-born fawn. A golden eagle, whose spring specialty is for fawns, kids and lambs, was seen to swoop swiftly down toward a solitary antelope that had been noticed on a treeless range beside the Little Missouri. It quickly became evident that the eagle was after an antelope fawn. As the bird swooped down toward the mother, and endeavored to seize her fawn in its talons, the doe rose high on her hind legs, and with her forelegs flying like flails struck with her sharp- pointed hoofs again and again. Her blows went home, and feathers were seen to fly from the body of the marauder.

The doe made good her defense. The eagle was glad to escape, and as quickly as possible pulled himself together and flew away.

The Defensive Circle of the Musk-Ox. Several arctic explorers have described the wonderful living-ring defense, previously mentioned, of musk-ox herds against wolves. Mr. Paul Rainey's moving pictures have shown it to us in thrilling detail, with Eskimo dogs instead of wolves. When a musk-ox herd is attacked by the big and deadly arctic white wolves, the bulls and adult cows herd the calves and young stock into a compact group, then take their places shoulder to shoulder around them in a perfect circle, and with lowered heads await the onset. The sharp down-and-up curved horn of the musk-ox is a deadly weapon against all the dangerous animals of the North, except man.

When a wolf approaches near and endeavors to make a breach in the circle, the musk-ox nearest him tries to get him, and will even rush out of the line for a short and brief pursuit. But the bull does not pursue more than twenty yards or so, for fear of being surrounded alone and cut off. At the end of his usually futile run, back he goes and carefully backs into his place in the first line of defense. A charging bull does not rush out far enough that the wolves can cut him off and kill him. He is much too wise for that.

Mr. Stefansson says that the impregnability of the musk-ox defense is so well recognized by the wolves of the North that often a pack will march past a herd in close proximity without offering to attack it, and without even troubling the herd to form the hollow circle.

A Savage Wild Boar. I once had a "fight" with a captive Japanese wild boar, under conditions both absurd and tragic, and from it I learned the courage and fury of such animals. The animal was large, powerful, fearfully savage toward every living thing, and insanely courageous. It was confined in a yard enclosed by a strong wire fence, and while we were all very sure that the fence would hold it, I became uneasy. In mid-afternoon I went alone to the spot, passing hundreds of school children on the way, to study the situation. When I reached the front of the corral and stood still to look at the fence, the boar immediately rushed for me. He came straight on, angry and terrible, and charged the wire like a living battering-ram. He repeated these charges until I became fearful of an outbreak, and decided to try to make him afraid to repeat them. Procuring from the bear dens, a pike pole with a stout spike in the end, I received the next charge with a return thrust meant to puncture both the boar's hide and his understanding. He backed off and charged more furiously than ever, with white foam flying from his jaws.

He cared nothing for his punishment. He charged until his snout bled freely, and the fence bulged at the strain.

Then I became regularly scared! I feared that the savage beast would break through the fence in spite of its strength, and run amuck among those helpless children. I "beat it" back to my office, hurried back with one of my loaded rifles, and without losing a second put a bullet through that raging brain and ended that danger forever.

The Overrated Peccary. This reminds me that the collared peccary has been credited with a degree of courage that has been much exaggerated. While a hunted and cornered peccary will fight dogs or men, and put up a savage and dangerous defense, men whom I know in the peccary belt of Mexico have assured me that a drove of peccaries will *not* attack a hunter who has killed one of their mates, nor keep him up a tree for hours while they swarm underneath him waiting for his blood. I have been assured by competent witnesses that in peccary hunting there is no danger whatever of mass attack through a desire for revenge, and that peccaries fired at will run like deer.

A Black Bear Killed a Man for Food. There is on record at least one well-authenticated case of a black bear deliberately going out of his way to cross a river, attack a man and kill him.

On May 17, 1907, at a lumber camp of the Red Deer Lumber Company, thirty miles south of Etiomami on the Canadian Northern Railway, Northwest Territory, a cook named T. Wilson was chased by a large black bear, without provocation, struck once on the head, and instantly killed. The bear then picked him up, carried him a short distance, and proceeded to *eat* him. Ten shots from a .32 calibre revolver had no effect. Later a rifle ball drove the bear away, but only after it had eaten the left thigh and part of the body. (Forest and Stream, Feb. 8, 1908.)

The Status of the Gray Wolf. In America wolves rarely succeed in killing men, although they often follow men's trails in the hope of spoil of some kind. But there are exceptions.

In 1912, around Lake Nipigon, Province of Ontario, Canada, there existed a reign of terror from wolves. The first man killed was a half-breed mail-carrier. Then, in December, another mail-carrier, who was working the lumber camps north of Lake Nipigon, was killed by wolves and completely devoured. The snow showed a terrible struggle, in which four large wolves had been killed by the carrier.

In Russia and in France in the days preceding the use of modern breech-loading firearms, the gray wolves of Europe were very bold, and a great many people were killed by them.

Killings by Wild Beasts in India. The killing by wild beasts of unarmed and defenseless native men, women and children in India is a very different matter from man-killing in resourceful and dangerous North America. The annual slaughter by wild beasts in Hindustan and British Burma is a fairly good index of the courage and aggressiveness of the parties of the first part. In India during the year 1878, in which we were specially interested, the totals were as follows:

Persons killed by elephants, 33; tigers, 816; leopards, 300; bears, 94; wolves, 845; hyenas, 33; snakes, 16,812.

Of course such slaughter as this by the ridiculous hyenas and the absurd sloth bears of India is possible only in a country wherein the swarming millions of people are universally defenseless, and children are superabundant.

As a corollary to the above figures, a comparison of them with the roster of wild animals killed and paid for is of some interest. The dangerous beasts destroyed were as follows:

Elephants, 1; tigers, 1,493; leopards, 3,387; bears, 1,283; wolves, 5,067; hyenas, 1,202; serpents, 117,782.

The Fighting Spirit in Baboons. In the first analysis, we find that courage is an individual trait, and that so far as we know, it never characterizes all the individuals of any one species. The strongest and the best armed of men and beasts usually are accounted the bravest ones of earth. The defenseless ones do well to be timid, to avoid hostilities and to flee from conflict to avoid being destroyed. It is just as much the duty of a professional mother to flee and to hide, in order to save her own life, as it is for "the old he-one" to threaten and to fight.

At the same time, there are many species which are concededly courageous, as species. In making up this list I would place first of all the baboons of eastern Africa, whom I regard collectively as the most bold and reckless fighters per pound avoirdupois to be found in the whole Order Primates. They have weapons, agility, strength and cyclonic courage. On no other basis could they have so long survived *on land* in a country full of lions, leopards, cheetahs, hyenas and wild dogs.

In order to appreciate the fighting spirit of a male baboon, the observer need only come just once in actual touch with one. A dozen times I have been seized by a powerful baboon hand shot out with lightning quickness between or under his cage bars. The combined strength and ferocity of the grab, and the grip on the human hand or arm, is unbelievable until felt, and this with an accompaniment of glaring eyes, snarling lips and nerve-ripping voice is quite sufficient to intimidate any ordinary man.

But even in the courage and belligerency of baboons, there are some marked differences between species. I rank them as follows:

The most fierce and dangerous species is the East African baboon.

The next for courage is the Rhodesian species.

The spectacular hamadryas baboon is a very good citizen. The long-armed yellow species makes very little trouble, and

The small golden baboon is the best-behaved of them all.

Courage in the Great Apes. After forty years of ape study, with many kinds of evidence, I am convinced that the courage and the alleged ferocity of the gorilla has been much over-rated. I believe this is due to the influence upon the human mind of the great size and terrifying aspect of the animal.

Of all the men whom I have known or read, the late R. L. Garner knew by far the most of gorilla habits and character by personal observation in the gorilla jungles of equatorial Africa. And never, in several years of intimate contact with Mr, Garner did he so much as once put forth a statement or an estimate that seemed to me exaggerated or overcolored.

In our many discussions of gorilla character Mr. Garner always represented that animal as very shy, wary of observation by man, profoundly cunning in raiding *in darkness* the banana plantations of man's villages, and most carefully avoiding exposures by daylight. He described the gorilla as practically never attacking men unless first attacked by them, and fleeing unless forcibly brought to bay.

He told me of are doubtable African tribesman who once captured a baby gorilla on the ground by suddenly attacking the mother with his club and beating her so successfully that she fled from him and abandoned her young. "But," said Mr. Garner, "there is only one tribe in Africa that could turn out a man who would attempt a feat like that."

That the gorilla can and will fight furiously and effectively when brought to bay is well known, and never denied.

Of the apes I have known in captivity, the chimpanzees are by far the most aggressive, courageous and dangerous. A vigorous male specimen over eight years of age is more dangerous than a lion, or tiger, or grizzly bear, and *far more anxious* to fight something. I think that even if our Boma were muzzled, no five men of my acquaintance could catch him and tie his hands and feet.

The orang-utan is only half the fighter that the chimpanzee is. Even the adult males are not persistently aggressive, or inflamed by savage desires to hurt somebody.

Courage in Elephants as an Asset. In all portions of India wherein tiger hunting with elephants is practiced, elephants with good courage are at a premium. No elephant is fit to carry a howdah in a line of beaters, with a valuable sahib on board, unless its courage can stand the acid test of a wounded tiger's charge. When an elephant can endure without panic an infuriated tiger climbing up its frontispiece to get at the unhappy mahout and the hunter, that elephant belongs in the courageous class. The cowardly elephant screams in terror, bolts for the rear, and if there is a tree in the landscape promptly wrecks the howdah and the sportsman against its lower branches.

A "rogue" elephant always reminds me of my Barbados boatman's description of a pugnacious friend: "De trouble is, he am too brave!" A rogue elephant will attack anything from a wheelbarrow to a hut, and destroy it. The peak of rogue ambition was reached on a railway in Burma, near Ban Klap, in March 1908, when a rogue elephant "on hearing the locomotive whistle, trumpeted loudly and then, lowering his head, charged the oncoming train. The impact was tremendous. Such was the impetus of the great pachyderm that the engine was partially derailed, the front of the smoke-box shattered as far as the tubes, the cow-catcher was crushed into a shapeless piece of iron, and other damages of minor importance were sustained. The train was going thirty-four miles per hour, and the engine alone weighed between forty and fifty tons.

"Of course the elephant was killed by the shock, its head being completely smashed.... It is believed that this particular rogue had been responsible for considerable damage to villages in the vicinity of Lopbusi. A number of houses have been pulled down recently and havoc wrought in other ways."

On another occasion a vicious rogue elephant elected to try conclusions with a railway train. In 1906, on the Korat branch of the Siamese State Railway, a bull elephant attacked a freight train running at full speed. He charged the rushing locomotive, with the result that the locomotive and several cars were derailed and sent down the side of the grade, and two persons were killed. The elephant was killed outright and buried under the wreck of the train. This occurred in open country, where there was no excuse for an elephant on the track, and therefore the charge of the rogue was wholly gratuitous.

Captive elephants whose managers are too humane to punish them for manifestations of meanness become spoiled by their immunity, just as mean children are spoiled when fond and foolish parents feel that their little jackets are too sacred ever to be tanned. Such complete immunity is as bad for bad elephants as for bad children, but in practice the severe punishment of an elephant with real benefit to the animal is next door to an impossibility, and so we never attempt it. We do, however, inflict mild punishments, of the fourth order of efficiency.

Animals and Men. Among the animals that are most courageous against man are the species and individuals that are most familiar with him, and feel for him both contempt and hatred. The cat scratches, the bad dog bites, the vicious horse kicks or bites, and the mean pet bear, tiger, ape, leopard, bison or deer will attempt injury or murder whenever they think the chance has arrived. I know a lady whose pet monkey is a savage and mean little beast, and because she never thrashes it as it deserves, both of her arms from wrist to elbow have been scarified by its teeth.

Mr. E. R. Sanborn, official photographer of the Zoological Park, once made an ingenious and also terrifying experiment. He made an excellent dummy keeper, stood it up, and tied it fast against the fence inside the yard of our very large and savage male Grevy Zebra. Then he posed his moving picture camera in a safe place, and the keeper turned the zebra into the yard. The moment the bad zebra caught sight of the presumptive keeper,—at last within his power,—he rushed at the dummy with glaring eyes and open mouth, and seized his victim by the head. With furious efforts he tore the dummy loose from its moorings, whirled it into the middle of the yard, where in a towering rage he knelt upon it, bit and tore its heart out. Of course the unfortunate dummy perished. The zebra reveled in his triumph, and altogether it was a fearsome sight.

CAUTION. A thoroughly cowardly horse *never* should be ridden, nor driven to anything so light that a runaway is possible. Such animals are too expensive both to human life and to property. A dangerous horse can be just as great a risk as a bad lion or bear.

IV.—THE BASER PASSIONS

XXII

FEAR AS A RULING PASSION

If we were asked, "Which one may be called the ruling passion of the wild animal?" we would without hesitation answer,—it is fear.

From the cradle to the grave, every strictly wild animal lives, day and night, in a state of fear of bodily harm, and dread of hunger and famine.

"Now the 'free, wild life' is a round of strife, And of ceaseless hunger and fear; And the life in the wild of the animal child Is not all skittles and beer."

The first thing that the wild baby learns, both by precept and example, is safety first! When the squalling and toddling bear cub first goes abroad, the mother bear is worried and nervous for fear that in a sudden and dangerous emergency the half-helpless little one will not be able to make a successful get-away when the alarm-signal snort is given. During the first, and most dangerous, days in the life of the elk, deer and antelope fawn, the first care of the mother is to hide her offspring in a spot cunningly chosen beside a rock, beside a log, or in thick bushes. In the absence of all those she looks for a depression in the earth wherein the fawn can lie without making a hump in the landscape. The first impulse of the fawn, — even before nursing if the birth occurs in daylight, — is to fold its long legs, short body and reptilian neck into a very small package, hug the earth tightly, close its eyes and lie absolutely motionless until its mother gives the signal to arise and sup. Such infants may lie for long and weary hours without so much as moving an ear; and the anxious mother strolls away to some distance to avoid disclosing her helpless offspring.

Now, suppose you discover and touch an elk or a deer fawn while thus hiding. What will it do? Nine times out of ten it will bound up as if propelled by steel springs, and go off like an arrow from a bow, dashing in any direction that is open and leads straight away. The horrified mother will rush into view in dangerously near proximity, and I have seen a wild white-tailed deer doe tear madly up and down in full view and near by, to attract the danger to herself.

Thousands of men and boys have seen a mother quail flop and flutter and play wounded, to lead the dangerous boy away from her brood of little quail mites, and work the ruse so daringly and successfully as to save both her babies and herself. I well remember my surprise and admiration when a mother quail first played that trick upon me. I expected to pick her up, — and forgot all about the chicks, — until they were every one safely in hiding, and then Mrs. Quail gave me the laugh and flew away.

Was it strategy? Was it the result of quail thought and reason? Or did it come by heredity, just like walking? To deny the cold facts in the quail case is to discredit our own ability to reason and be honest.

Fear is the ruling emotion alike of the most timid creatures, and also the boldest. Of course each wild animal keeps a mental list of the other animals of which he is not afraid; and the predatory animal also keeps a card catalogue of those which he may safely attack when in need of food.

But, with all due consideration to mighty forearm, to deadly claws and stabbing fangs, there is (I think) absolutely no land animal that is not afraid of something. Let us progressively consider a few famous species near at hand.

The savage and merciless weasel fears the fox, the skunk, the wolf and the owl. The skunk fears the coyote which joyously kills him and devours all of him save his jaws and his tail. The marten, mink and fisher have mighty good reason to fear the wolverine, who in his turn cheerfully gives the road to the gray wolf. The wolf and the lynx carefully avoid the mountain lion and the black bear, and the black bear is careful not to get too close to a grizzly. Today a cottontail rabbit is not more afraid of a hound than a grizzly bear is of a man. The polar bear once was bold in the presence of man; but somebody has told him about breech-loading high power rifles; and now he, too, runs in terror from every man that he sees. The lion, the tiger, the leopard and the jaguar all live in wholesome fear of man, and flee from him at sight. The lordly elephant does likewise, and so does the rhinoceros, save when he is in doubt about the identity of the biped animal and trots up to get certainty out of a nearer view. Col. Roosevelt became convinced, that most of the alleged "charging" of rhinoceroses was due to curiosity and poor vision, and the desire of rhinos to investigate at close range.

Today the giant brown bears of Alaska exhibit less fear of man than any other land animals that we know, and many individuals have put themselves on record as dangerous fighters. And this opens the door to the great Alaskan controversy that for a year raged, —chiefly upon one side, —in certain Alaskan newspapers and letters.

Early in 1920, certain parties in Alaska publicly asked people to believe that W. T. Hornaday in his "published works" had set up the Alaskan brown bear as "a harmless animal." All these statements and insinuations were notoriously false, but the repetition of them went on right merrily, even while the author's article portraying the savage and dangerous character of the brown bear was being widely circulated in the United States through *Boys' Life* magazine.

The indisputable facts regarding the temper of the great Alaskan brown bears are as follows: Usually, unless fired at, these big brown bears flee from man at sight of him, and by many experienced Alaskan bear hunters who can shoot they are not regarded as particularly dangerous, save when they are attacked by man, or think that they are to be attacked.

They are just now the boldest of all bears, and the most dangerous.

They often attack men who are hunting them, and have killed several.

They have attacked a few persons who were not hunting.

Where they are really numerous they are a menace and a nuisance to frontiersmen who need to traverse their haunts.

In all places where Alaskan brown bears are quite too numerous for public safety, their numbers should thoroughly be reduced; and everywhere the bears of Alaska should be pursued and shot until the survivors acquire the wholesome respect for man that now is felt everywhere by the polar and the grizzly. Then the Alaskans will have peace, and our Alaskan enemies possibly will cease to try to discredit our intelligence.

The most impressive exhibition of wild-animal fear that Americans ever have seen was furnished by the African motion pictures of Paul J. Rainey. They were taken from a blind constructed within close range of a dry river bed in northern British East Africa, where a supply of water was held, by a stratum of waterproof clay or rock, about four feet below the surface of the dry river bed. By industrious pawing the zebras had dug a hole down to the water, and to this one life-saving well wild animals of many species flocked from miles around. The camera faithfully recorded the doings of elephants, giraffes, zebras, hartebeests, gnus, antelopes of several species, wart-hogs and baboons.

The personnel of the daily assemblage was fairly astounding, and to a certain extent the observer of those wonderful pictures can from them read many of the thoughts of the animals.

Next to the plainly expressed desire to quench their thirst, the dominant thought in the minds of those animals, one and all, was the *fear of being attacked.* In some species this ever- present and harassing dread was a pitiful spectacle. I wish it might be witnessed by all those ultra-humane persons who think and say that the free wild animals are the only happy ones!

With the possible exception of the sanguine-tempered elephants, all those animals were afraid of being seized or attacked while drinking. One and all did the same thing. An animal would approach the water-hole, nervously looking about for enemies. The fore feet cautiously stepped down, the head disappeared to reach the water, —but quickly shot upward again, to look for the enemies. It was alternately drink, look, drink, look, for a dozen quick repetitions, then a scurry for safety.

Even the stilt-legged and long-necked giraffes went through that same process,—a mouthful of water greedily seized, and a fling of the head upward to stare about for danger. Group by group the animals of each species took their turns. The baboons drifted down over the steep rocky slope like a flock of skimming birds, and watched and drank by turn. Having finished, they paused not for idle gossip or play, but as swiftly as they came drifted up the slope and sought safety elsewhere.

And yet, it was noticeable that during the whole of that astounding panorama of ferae naturae unalloyed by man's baleful influence, no species attacked another, there was no fighting, nor even any threatening of any kind. Had there been a white flag waving over that water-hole, the truce of the wild could not have been more perfect.

Effect of Fear in Captive Animals. Among captive wild animals, by far the most troublesome are those that are obsessed by slavish fear of being harmed. The courageous and supremely confident grizzly or Alaskan brown bear is in his den a good-natured and reliable animal, who obeys orders when the keepers enter the den to do the daily housework and order him to "Get up out of here." The fear-possessed Japanese black bear, Malay sun bear and Indian sloth bear are the ones that are most dangerous, and that sometimes charge the keepers.

Our famous "picture lion," Sultan, was serenely confident of his own powers, his nerves were steady and reliable, and he never cared to attack man or beast. Once when by the error of a fellow keeper the wrong chain was pulled, and the wrong partition door was opened, the working keeper bent his head, and broom in hand walked into what he thought was an empty cage. To his horror, he found himself face to face with Sultan, with only the length of the broom handle between them.

The startled and helpless keeper stood still, and said in a calm voice, without batting an eye.

"Hello, Sultan."

Sultan calmly looked at him, wonderingly and inquiringly, but without even a trace of excitement; and feeling sure that the keeper did not mean to harm him, he seemed to have no thought of attacking.

The keeper quietly backed through the low doorway, and gently closed the door. Had the keeper lost his nerve, *and shown it,* there might have been a tragedy.

Lions are the best of all carnivorous performing animals, because of their courage, serenity, self-confidence and absence of jumpy nerves. Leopards are the worst, and polar bears stand next, with big chimpanzees as a sure third. Beware of all three.

Exceptions to the Rule of Fear. Fortunately for the wild animal world, there are some exceptions to the rule of fear. I will indicate the kinds of them, and students can supply the individual cases.

Whenever a wild animal species inhabits a spot so remote and inaccessible that man's blighting hand never has fallen upon it, nor in any way influenced its life or its fortunes, that species knows no fear save from the warring elements, and from predatory animals. The wonderful giant penguins found and photographed near the south pole by Sir Ernest Shackleton never had seen nor heard of men, never had been attacked by predatory animals or birds. You may search this wide world over, and you will not find a more striking example of sublime isolation. Those penguins had been living in a penguin's paradise. The sea-leopard seals harmed them not, and until the arrival of the irrepressible British explorer the spell of that antarctic elysium was unbroken.

[Illustration
with caption: PRIMITIVE PENGUINS ON THE ANTARCTIC CONTINENT,
UNAFRAID OF MAN (From Sir Ernest Shackleton's "Heart of the Antarctic," by permission of William Heinemann and the J. B. Lippincott Company, publishers)]

Those astounding birds knew no such emotion as fear. Under the impulse of the icy waves dashing straight up to the edge of the ice floes, those giant penguins shot out of the water, sped like catapulted birds curving through the air, and landed on their cushioned breasts high and dry, fully ten feet back from the edge of the floe. They flocked together, they waddled about erect and serene, heads high in air, and marched close up to the ice-bound ship to see what it was all about. Men and horses freely walked among them without exciting fear, and when the birds gathered in a vast assemblage the naturalists and photographers were welcomed everywhere.

And indeed those birds were well-nigh the most fortunate birds in all the world. The men who found them were not low-browed butchers thinking only of "oil" or "fertilizer"; and they did not go to work at once to club all those helpless birds into masses of death and corruption. Those men wondered at them, laughed at them, photographed them, studied them, — and *left them in peace!*

What a thundering contrast that was with the usual course of Man, the bloody savage, under such circumstances! The coast of Lower California once swarmed with seals, sea-lions and birds, and the waters of the Gulf were alive with whales. Now the Gulf and the shores of the Peninsula are as barren of wild life as Death Valley.

The history of the whaling industry contains many sickening records of the wholesale slaughter by savage whalers of newly discovered herds of walrus, seals and sea birds that through isolation knew no fear, and were easily clubbed to death en masse.

Wild creatures generally subscribe to the political principle that in union there is strength. In the minds of wild animals, birds and reptiles, great numbers of individuals massed together make for general security from predatory attacks. The herd with its many eyes and ears feels far greater security, and less harrowing fear, than the solitary individual who must depend upon his own two pair. The herd members relax and enjoy life; but the solitary bear, deer, sheep, goat or elephant does not. His nerves always are strung up to concert pitch, and while he feeds or drinks, or travels, he watches his step. A moving object, a strange-looking object, a strange sound or a queer scent in the air instantly fixes his attention, and demands analysis.

On the North American continent the paramount fear of the wild animal is aroused to its highest pitch by what is called "man scent." And really, from the Battery to the North Pole, there is good reason for this feeling of terror, and high wisdom in fleeing fast and far.

Said a wise old Ojibway Indian to Arthur Heming:

"My son, when I smell some men, and especially some white men, I never blame the animals of the Strong Woods for taking fright and running away!"

And civilization also has its terrors, as much as the wilderness.

The fox, no matter what is the color of his coat, or his given name, is the incarnation of timidity and hourly fear. The nocturnal animals go abroad and work at night solely because they are afraid to work in the daytime. The beaver will cheerfully work in daytime if there is no prospect of observation or interference by man. The eagle builds in the top of the tallest tree, and the California condor high up on the precipitous side of a frightful canyon wall, because they are afraid of the things on the ground below. In the great and beautiful Animallai Forest (of Southern India), in 1877 the tiger walked abroad in the daytime, because men were few and weak, but in the populous and dangerous plains he did his traveling and killing at night, and lay closely hidden by day.

Judging by the records of those who have hunted lions, I think that naturally the lion has more courage and less fear of bodily harm than any other wild animal of equal intelligence. By reason of his courage and self-confidence, as well as his majesty of physique, the lion is indeed well worthy to be called the King of Beasts.

Among the few animals that seem naturally bold and ready to take risks, a notable species is the gray wolf. But is it really free from fear? Far from it. When in touch with civilization, from dawn until dark the wolf never forgets to look out for his own safety. He fears man, he fears the claws of every bear, he fears traps, poison and the sharp horns of the musk-ox. Individually the wolf is a contemptible coward. Rarely does he attack all alone an animal of his own size, unless it is a defenseless colt, calf or sheep. No animal is more safe from another than an able-bodied bull from the largest wolf. The wolf believes in mass action, not in single combat.

But there is hope for the harassed and nerve-racked children of the wild. *The Game Sanctuary has come!* Its area of safety, and its magic boundary, are quickly recognized by the harried deer, elk, sheep, goat and antelope, and right quickly do these and all other wild animals set up housekeeping on a basis of absolute safety. Talk about wild animals not "reasoning!" For shame. What else than REASON convinced the wild mountain sheep in the rocky fastnesses they once inhabited in terror that now they are SAFE, even in the streets of Ouray, and that "Ouray" rhymes with "your hay"?

On account of his crimes against wild life, man (both civilized and savage) has much to answer for; but each wild life sanctuary that he now creates wipes out one chapter. From the Cape to Cairo, from the Aru Islands to Tasmania and from Banks Land to the Mexican boundary, they are growing and spreading. In them, save for the misdoings of the few uncaught and unkilled predatory animals, fear can die out, and the peace of paradise regained take its place.

HYSTERIA OF FEAR IN A BEAR. Among wild animals in captivity hysteria, of the type produced by fear, is fairly common. A case noticed particularly on October 16, 1909, in a young female Kadiak bear, may well be cited as an example.

The subject was then about two and one-half years old, and was caged in a large open den with four other bears of the same age. Of a European brown bear male, only a trifle larger than herself, she elected to be terror-stricken, as much so as ever a human child was in terror of every move of a brutal adult tormentor. Strangely enough, the cause of all this terror was wholly unconscious of it, and in the course of an observation lasting at least twenty minutes he made not one hostile movement. The greater portion of the time he idly moved about in the central space of the den, wholly oblivious of the alarm he was causing.

The young Kadiak, in full flesh and vigor, first attracted my attention by her angry and terrified snorting, three quick snorts to the series. On the top of the rocks she raced to and fro, constantly eyeing the bear in the centre of the den. If he moved toward the rocks, she wildly plunged down, snorting and glaring, and raced to the front end of the den. If the bogey stopped to lick up a fallen leaf, she took it as a hostile act and wildly rushed past him and scrambled up the rocks at the farther end of the den. This was repeated about fifteen times in twenty minutes, accompanied by a continuous series of terrified snorts. She panted from exhaustion, frothed at the mouth, and acted like an animal half crazed by terror.

Not once, however, did the bogey bear pay the slightest attention to her, and his sleepy manner was anything but terrifying.

These spells of hysteria (without real cause) at last became so frequent that they seemed likely to injure the growth of a valuable animal, and finally the bogey bear was removed to another den.

XXIII

FIGHTING AMONG WILD ANIMALS

Quarrels and combats between wild animals in a state of nature are almost invariably due to one of two causes—attack and defense in a struggle for prey, or the jealousy of males during the mating season. With rare exceptions, battles of the former class occur between animals of different Orders,—teeth and claws against horns and hoofs, for instance; and it is a fight to the death. Hunger forces the aggressor to attack something, and the intended victim fights because it is attacked. The question of good or ill temper does not enter in. On both sides it is a case of "must," and neither party has any option. Such combats are tests of agility, strength, and staying powers, and, in a few cases, of thickness of bone and hide.

How Orang-Utans Fight. Of the comparatively few animals which do draw blood of their own kind through ill temper or jealousy, I have never encountered any more given to internecine strife than orang-utans. Their fighting methods, and their love of fighting, are highly suggestive of the temper and actions of the human tough. They fight by biting, and usually it is the fingers and toes that suffer. Of twenty-seven orang-utans I shot in Borneo, and twelve more that were shot for me by native hunters, five were fighters, and had had one or more fingers or toes bitten off in battle. Those specimens were taken in the days when the museums of America were one and all destitute of anthropoid apes.

A gorilla, chimpanzee, or orang-utan, being heavy of body, short of neck, and by no means nimble footed, cannot spring upon an adversary, choose a vulnerable spot, and bite to kill; but what it lacks in agility it makes up in length and strength of arm and hand. It seizes its antagonist's hand, carries it to its own mouth, and bites at the fingers. Usually, the bitten finger is severed as evenly as by a surgeon's amputation, and heals quite as successfully.

I never saw two big orang-utans fighting, but I have had several captive ones seize my arm and try to bring my fingers within biting distance. The canine teeth of a full grown male orang, standing four feet four inches in height, and weighing a hundred and fifty pounds or more, are just as large and dangerous as the teeth of a bear of the same size, and the powerful incisors have one quality which the teeth of a bear do not possess. A bear pierces or tears an antagonist with his canines, but very rarely bites off anything. An orang-utan bites off a finger as evenly as a boy nips off the end of a stick of candy.

When orang-utans fight, they also attack each other's faces, and often their broad and expansive lips suffer severely. My eleventh orang bore the scars of many a fierce duel in the tree-tops. A piece had been bitten out of the middle of both his lips, leaving in each a large, ragged notch. Both his middle fingers had been taken off at the second joint, and his feet had lost the third right toe, the fourth left toe, and the end of one hallux. His back, also, had sustained a severe injury, which had retarded his growth. This animal we called "The Desperado."

Orang No. 34 had lost the entire edge of his upper lip. It had been bitten across diagonally, but adhered at one corner, and healed without sloughing off, so that during the last years of his life a piece of lip two inches long hung dangling at the corner of his mouth. He had also suffered the loss of an entire finger. No. 36 had lost a good sized piece out of his upper lip, and the first toe had been bitten off his left foot.

All these combats must have taken place in the tree-tops, for an adult orang-utan has never been known to descend to the earth except for water. In some manner it has become a prevalent belief that in their native jungles all three of the great apes— gorilla, orang, and chimpanzee—are dangerous to human beings, and often attack them with clubs. Nothing could be farther from the truth. According to the natives of West Africa, a gorilla or chimpanzee fights a hunter by biting his face and fingers, just as an orang-utan does. I believe that no sane orang ever voluntarily left the safety of a tree top to fight at a serious disadvantage on the ground; and I am sure an orang never struck a blow with a club, unless carefully taught to do so.

WILD ANIMALS ARE NOT QUARRELSOME. As a species, man appears to be the most quarrelsome animal on the earth; and the same quality is strongly reflected in his most impressionable servant and companion, the domestic dog. Nearly all species of wild animals have learned the two foundation facts of the philosophy of life— that peace is better than war, and that if one must fight, it is better to fight outside one's own species. To this rule, however, wolves are a notable exception; for wherever wolves are abundant a wounded wolf is a subject for attack, and usually it is killed and eaten by the other members of the pack.

I have observed the daily habits of many kinds of wild animals in their wild haunts, but in the field I never yet have seen either a fight between animals of the same species, or between two of different species. This may seem a very humiliating admission for a hunter to make, but it happens to be true. In the matter of finding big snakes, having exciting adventures, and witnessing combats between wild animals, there are some men who never are in luck.

Now there was the "Old Shekarry,"—whose elephants, tigers, bison, bears, and sambar always were so much larger than mine. In his book, "Sport in Many Lands," he describes an affair of honor between a tiger and a bull bison, which was a truly ideal combat. The champions met by appointment,—by the light of the moon, in order to be safe from interference by the jungle police,—and they fought round after round, in the most orthodox prize ring style, under the Queensberry rules. So fairly did they fight that neither claimed a foul, and at the finish the two combatants retired to their respective corners and died simultaneously, "to the musical twitter of the night bird."

Another writer has given a vivid description of a battle to the death between a wild bull and a grizzly bear; and we have read of several awful combats between black bears and alligators, in Florida; but some of us have yet to find either a black bear or an alligator that will stop to fight when he has an option on a line of retreat. When he has lived long,—say to the length of twelve feet,—the alligator is a hideous and terrorizing beast; but, for all that, he knows a thing or two; and a full grown, healthy black bear of active habit is about the last creature on earth that a 'gator would care to meddle with. Pigs and calves, fawns, stray dogs, ducks and mud hens are antagonists more to his liking.

The Fighting Tactics of Bears. In captivity, bears quarrel and scold one another freely, at feeding time, but seldom draw blood. I have questioned many old hunters, and read many books by bear hunters, but Ira Dodge, of Wyoming, is the only man I know who has witnessed a real fight between wild bears. He once saw a battle between a cinnamon and a grizzly over the carcass of an elk.

In attacking, a bear does three things, and usually in the same order. First, he delivers a sweeping sidewise blow on the head of his antagonist; then he seizes him by the cheek, with the intention of shifting to the throat as quickly as it is safe to do so. His third move consists in throwing his weight upon his foe and bearing him to the earth, where he will have a better chance at his throat. If the fighters are fairly matched, the struggle is head to head and mouth to mouth. After the first onset, the paws do little or no damage, and the attacks of the teeth rarely go as far down as the shoulders. Often the assailant will seize his opponent's cheek and hold on so firmly that for a full minute the other can do nothing; but this means little.

In combats between bears, the one that is getting mauled, or that feels outclassed, will throw himself upon the ground, flat upon his back, and proceed to fight with all four sets of claws in addition to his teeth. This attitude is purely defensive, and often is maintained until an opportunity occurs to attack with good advantage, or to escape. It is very difficult for a standing bear to make a serious impression upon an antagonist who lies upon his back, clawing vigorously with all four feet at the head of his assailant.

Tiger Versus Grizzly Bear. Often is the question asked, "If a grizzly bear and a tiger should fight, which would whip the other?" One can answer only with opinions and deductions, not by reference to the records of the ring; for it seems that the terrors of the occident and the orient have never yet been matched in a fight to a finish.

One of the heaviest tigers ever weighed, prior to 1878, scaled four hundred and ninety five pounds, and was as free from surplus flesh and fat as a prizefighter in the ring. He stood three feet seven inches at the shoulder, measured thirty-six inches around the jaws, and twenty inches around the forearm. Very few lions have ever exceeded his weight or dimensions. So far as I know, a wild grizzly bear of the largest size has never been scaled, but it is not at all certain that any California grizzly has weighed more than twelve hundred pounds. The silvertip of the Rocky Mountain region is a totally different animal, being smaller, as well as different in color.

In a match between a grizzly and a tiger of equal weights, the activity of the latter, combined with the greater spread of his jaws and length of his canine teeth, would insure him the victory. The superior attack of the tiger would give him an advantage which it would probably be impossible to overcome. The blow of a tiger's paw is as powerful as that of a grizzly of the same size, though I doubt if it is any quicker in delivery. The quickness with which a seemingly clumsy bear can deliver a smashing blow is astonishing. Moreover, nature has given the grizzly a coat of fur which as a protection in fighting is almost equal to chain mail. Its length, combined with its density, makes it difficult for teeth or claws to cut through it, and in a struggle with a tiger, protective fur is only a fair compensation for a serious lack of leaping power in the hinder limbs. Though the tiger would win at equal weights, it is extremely probable that an adult California grizzly would vanquish a tiger of the largest size, for his greater bulk would far outweigh the latter's agility.

The Great Cats as Fighters. Tigers, when well matched, fight head to head and mouth to mouth, as do nearly all other carnivora, and at the same time they strike with their front paws. One of the finest spectacles I ever witnessed was a pitched battle between two splendid tigers, in a cage which afforded them ample room. With loud, roaring coughs, they sprang together, ears laid tight to their heads, eyes closed until only sparks of green and yellow fire flashed through four narrow slits, and their upper lips snarling high up to clear the glittering fangs beneath. Coughing, snarling, and often roaring furiously, each sprang for the other's throat, but jaw met jaw until their teeth almost cracked together. They rose fully erect on their hind legs, with their heads seven feet high, stood there, and smashed away with their paws, while tufts of hair flew through the air, and the cage seemed full of sparks. Neither gave the other a chance to get the throat hold, nor indeed to do aught else than ward off calamity; and each face was a picture of fury.

This startling combat lasted a surprisingly long time, without noticeable advantage to either side. Finally the tigers backed away from each other, and when at a safe distance apart dropped their front feet to the floor, growling savagely and licking their lips wherever a claw had drawn blood.

Of all the wild animals that are preyed upon by lions, tigers, leopards, jaguars, and pumas, only half a dozen species do anything more than struggle to escape. The gaur and the wild buffalo of India are sufficiently vindictive in dealing with a human hunter whose aim is not straight, but both fly before the tiger, and count themselves lucky when they can escape with nothing worse to show than a collection of long slits on their sides and hind quarters made by his knife-like claws. They do not care to return to do battle for the sake of revenge, and seek to put the widest possible stretch of jungle between themselves and their dreaded enemy.

The same is true of the African buffalo and the lion. As to the antelopes of Africa and the deer of India, what can they do but make a desperate effort to escape, and fly like the wind whenever they succeed? Of course many of these defenseless animals make a gallant struggle for their lives, and not a few succeed in throwing off their assailants and escaping. Even domestic cattle sometimes return to the hill country villages of southern India bearing claw marks on their sides—usually the work of young tigers, or of rheumatic old ones.

Here is a deer and puma story. In the picturesque bad-lands of Hell Creek, Montana, I saw my comrade, Laton A. Huffman, kill a large mule deer buck that three months previously had been attacked by a puma. From above it, the great cat had leaped upon the back of the deer, and laid hold with teeth and claws. In its struggle for life the buck either leaped or fell off the edge of a perpendicular "cut bank," and landed upon its back, with the puma underneath. Evidently the puma was so seriously injured that it could not continue the struggle; but it surely left its ear-marks.

One ear of the buck was fearfully torn. There was a big wound on the top of the neck, where the puma jaws had lacerated the skin and flesh; and both hind legs had been badly clawed by the assailant's hind feet. The main beam of the right antler had been, broken off half-way up, while the antlers were still in the velvet, which enabled us to fix the probable date of the encounter.

In the great Wynaad forest I once got lost, and in toiling through a five acre patch of grass higher than my head, and so dense that it was not negotiable except by following the game trails, my simple old Kuramber and I came suddenly upon the scene of a great struggle. In the center of a space about twenty feet in diameter, on which the tall grass had been trampled flat, lay the remains of a sambar stag which had very recently been killed and eaten by a tiger. The neck had not been dislocated, and the sambar had fought long and hard. Evidently the tiger had lain in wait on the runway, and had failed to subdue the sambar by his first fierce onslaught. Now an angry stag with good antlers is no mean antagonist, and it is strange if the tiger in the case went through that struggle without a puncture in his tawny skin.

In South Africa, Vaughan Kirby once found the dead bodies of a "patriarchal bull" sable antelope and a lion, "which had evidently been a fine specimen," lying close together, where the two animals had fallen after a great struggle. The sable antelope must have killed its antagonist by a lucky backward thrust of its long, curved horns as the lion fastened upon its back to pull it down.

Mr. Kirby's dogs once disturbed a sanguinary struggle between a leopard and a wild boar, or "bush pig," which had well-nigh reached a finish. The old boar, when bayed by the dogs, was found to be most terribly mauled. Its tough skin hung literally in shreds from its neck and shoulders, presenting ghastly open wounds. The entrails protruded from a deep claw gash in the side, and the head was a mass of blood and dirt. "On searching around," says Mr. Kirby, "we found unmistakable evidence of a life and death struggle. The ground was covered with gouts of blood and yellow hair, to some of which the skin (of the leopard) was still attached. Blood was splashed plentifully on the tree stems and the low brushwood, which for a space of a dozen yards around was trampled flat." The leopard had fled upon the approach of the dogs, leaving a trail of blood, which, though followed quickly, was finally lost in bad ground. It is no wonder that from the above and many other evidences equally good, Mr. Kirby considers the bush pig a remarkably courageous animal. He says that it was "never yet known to show the white feather," and declares that "a pig is never defeated until he is dead."

The Combats of Male Deer. The sable antelope is one of the few exceptions to the well-nigh universal rule against fighting between wild animals of the same species. Of this species, Mr. Kirby says: "Sable antelope bulls fight most fiercely amongst themselves, and though I have never actually witnessed an encounter between them, I have often seen the results of such, evidenced by great gaping wounds that could have been made by nothing else than the horns of an opponent. I once killed a large bull with a piece of another's horn tip, fully three inches long, buried in its neck. In 1889 I shot an old bull on the Swinya with a terrible wound in its off shoulder, caused by a horn thrust."

During the jealous flashes of the mating season, the males of several species of deer fight savagely. After a long period of inaction while the new antlers are developing—from April to September—the beginning of October finds the male deer, elk, or moose of North America with a new suit of hair, new horns, a swollen neck, and all his usual assertiveness. The crisp autumn air promotes a disposition to fight something, precisely as it inspires a sportsman to "kill something." During October and November, particularly, it is well for an unarmed man to give every antlered deer a wide berth.

At this period, fights between the males of herds of mule deer, white-tailed deer and elk are of frequent occurrence, but in a wild state they rarely end in bloodshed or death, save from locked antlers. Many times, however, two bucks will come together, and playfully push each other about without being angry. Many pairs of bucks have been found with their antlers fast locked in death—and I never see a death lock without a feeling of grim satisfaction that neither of the quarrelsome brutes had had an opportunity to attack some defenseless man, and spear him to death.

The antlers of the common white-tailed deer seem peculiarly liable to become interlocked so tightly that it is well-nigh impossible to separate them. And whenever this happens, the doom of both deer is sealed. Unless found speedily and killed, they must die of starvation. While it is quite true that two deer playing with their antlers may become locked fast, it is safe to say that the great majority meet their fate by charging each other with force sufficient to spring the beams of their antlers, and make the lock so perfect that no force they can exert will release it. A deer cannot pull back with the same power it exerts in plunging forward.

All members of the deer family that I know follow the same natural law in regard to supremacy. Indeed, this is true of nearly all animals. Leadership is not always maintained by the largest and strongest member of a herd, but very often by the most pugnacious. Sometimes a herd of elk is completely tyrannized by an old doe, who makes the young bucks fly from her in terror, when one prod of their sharp antlers would quickly send her to the rear.

When bucks in a state of freedom fight for supremacy, the weaker does not stay to be overthrown and speared to death by the victor. As soon as he feels that he is mastered he releases his antlers at the first opportunity, flings himself to one side, and either remains in the herd as an acknowledged subject of the victor, or else seeks fresh fields and pastures new.

Battles in Zoological Parks. In captivity, where escape is impossible, it is no uncommon thing for elk to kill each other. In fact, with several adult males in a small enclosure, tragedies may always be expected in the autumn and early winter. The process is very simple. So long as the two elk can stand up and fight head to head, there are no casualties; but when one wearies and weakens before the other, its guard is broken. Then one strong thrust in its side or shoulder sends it to the earth, badly wounded; and before it can rise, it is generally stabbed to death with horn thrusts into its lungs and liver. But, as I said before, I have never known of a fatal duel between elk outside of a zoological garden or park.

One of the most novel and interesting fights that has yet taken place in the New York Zoological Park was a pitched battle between two cow elk—May Queen and the Dowager. A bunch of black fungus suddenly appeared on the trunk of a tree, about twelve feet from the ground. My attention was first called to this by seeing May Queen, a fine young cow, standing erect on her hind legs in order to reach the tempting morsel with her mouth. A little later the Dowager, the oldest and largest cow elk in the herd, met her under the tree, whereupon the two made wry faces at each other, and champed their teeth together threateningly. Suddenly both cows rose on their hind legs, struck out viciously with their sharp pointed front hoofs, and, after a lively sparring bout, they actually clinched. The young cow got both front legs of the old cow between her own, where they were held practically helpless, and then with her own front hoofs she fiercely rained blows upon the ribs of her assailant. The Dowager backed away and fled, completely vanquished, with May Queen close upon her heels; and thus was the tyrannical rule of the senior cow overthrown forever.

During the breeding season, our wild buffaloes of the great vanished herds were much given to fighting, and always through jealousy. The bulls bellowed until they could be heard for miles, tore up earth and threw it into the air, rolled their eyes, and often rushed together in a terrifying manner; but beyond butting their heads, pushing and straining until the weaker turned and ran, nothing came of it all. I have yet to find a man who ever saw a wild buffalo that had been wounded to the shedding of blood by another wild buffalo. It is probable that no other species ever fought so fiercely and did so little damage as the American bison.

Elephants, Wolves, and Others. In ordinary life the Indian elephant is one of the most even-tempered of all animals. I have spent hours in watching wild herds in southern India, sometimes finding the huge beasts all around me, and in dangerously close proximity. Several times I could have touched a wild elephant with a carriage whip, had I possessed one. So far from fighting, I never saw an elephant threaten or even annoy another.

Elephants, being the most intelligent of all animals in the matter of training, have been educated to fight in the arena, usually by pushing each other head to head. A fighting tusker can lord it over almost any number of tuskless elephants, because he can pierce their vitals, and they cannot pierce his. A female fights by hitting with her head, striking her antagonist amidships, if possible. Once when the late G. P. Sanderson was in a keddah, noosing wild elephants, and was assulted [sic] by a vicious tusker, his life was saved by a tame female elephant, whose boy driver caused her to attack the tusker with her head, and nearly bowl him over by the force of her blows upon his ribs.

In captivity, wolves are the meanest brutes on earth, and in a wild state they are no better. As a rule, the stronger ones are ever ready to kill the weaker ones, and eat them, too. One night, our male Russian wolf killed his mate, and ate nearly half of her before morning. A fox or a wolf cub which thrusts one of its legs between the partition bars and into a wolf's den almost invariably gets it bitten off as close to the body as the biter can go. In the arctic regions, north of the Great Slave Lake, "Buffalo" Jones and George Rea fought wolves incessantly for several days, and every wolf they wounded was immediately killed and devoured by its pack mates.

In captivity, a large proportion of mammals fight, more or less; and the closer the confinement, the greater their nervousness and irritability, and the more fighting. Monkeys fight freely and frequently. Serpents, lizards, and alligators rarely do, although large alligators are prone to bite off the tails or legs of their small companions, or even to devour them whole. Storks, trumpeter swans, darters, jays, and some herons are so quarrelsome and dangerous that they must be kept well separated from other species, to prevent mutilation and murder. In 1900, when a pair of trumpeter swans were put upon a lake in Prospect Park, Brooklyn, with three brown pelicans for associates, they promptly assailed the pelicans, dug holes in their backs, and killed all three. The common red squirrel is a persistent fighter of the gray species, and, although inferior in size, nearly always wins.

A Fight Between a Whale and a Swordfish. One of the strangest wild animal combats on record was thus described in the Proceedings of the Zoological Society of London, for 1909.

"Mr. Malcolm Maclaren, through Mr. C. Davies Sherborn, F. Z. S., called the attention of the Fellows to an account of a fight between a whale and a swordfish observed by the crew of the fishing-boat 'Daisy' in the Hauraki Gulf, between Ponui Island and Coromandel, as reported in the 'Auckland Weekly News,' 19th Nov., 1908. A cow whale and her calf were attacked by a 12 ft. 6 in. swordfish, the object of the fish being the calf. The whale plunged about and struck in all directions with her flukes. Occasionally the fins of the swordfish were seen as he rose from a dive, his object apparently being to strike from below. For over a quarter of an hour the whale circled round her calf, lashing furiously and churning up the water so that the assailant was unable to secure a good opportunity for a thrust. At last, after a fruitless dive, the swordfish came close up and made a thrust at the calf, but received a blow from the whale's flukes across the back, which apparently paralyzed it. It was killed and hauled on board the boat without difficulty, while the whale and calf went off towards Coromandel with splashings and plungings. The whale's blow had almost knocked off the back fin of the swordfish, and heavily bruised the flesh around it. No threshers accompanied the swordfish."

Beyond question, as firearms and hunters multiply, all wild animals become more timid, less inclined to attack man, and also less inclined to attack one another. The higher creatures are the most affected by man's destructiveness of animal life, and the struggle for existence has become so keen that fighting for the glory of supremacy, or as a pastime, will soon have no important place in the lives of wild animals.

XXIV

WILD ANIMAL CRIMINALS AND CRIME

Many human beings are "good" because they never have been under the harrow of circumstances, nor sufficiently tempted to do wrong. It is only under the strain of strong temptation that human character is put through the thirty-third degree and tried out. No doubt a great many of us could be provoked to join a mob for murder, or forced to steal, or tortured into homicidal insanity. It is only under the artificial conditions of captivity, with loss of freedom, exemption from the daily fear of death, abundant food without compensating labor, and with every want supplied, that the latent wickedness of wild creatures comes to the surface. A captive animal often reveals traits never recognized in the free individual.

"Satan finds some mischief still for idle hands to do."

These manifestations are of many kinds; but we propose to consider the criminal tendencies of wild animals both free and captive.

The persistence of the mental and moral parallelism between men and wild animals is a source of constant surprise. In a state of freedom, untrammeled by anything save the fear of death by violence, the deer or the mountain sheep works out in his own way his chosen scheme for the survival of the fittest,—himself. In the wilds we see very few manifestations of the criminal instinct. A fight between wild elk bulls for the supremacy of a herd is not a manifestation of murder lust, but of obedience to the fundamental law of evolution that the largest, the strongest and the most courageous males of every herd shall do the breeding. The killing of natural prey for daily food is not murder. A starving wolf on the desolate barren grounds may even kill and devour a wounded pack-mate without becoming a criminal by that act alone. True, such a manifestation of hard-heartedness and bad taste is very reprehensible; but its cause is hunger, not sheer blackness of heart. Among wild animals, the wanton killing of a member of the killer's own species would constitute murder in the first degree, and so is all unnecessary and wanton killing outside the killer's own species.

To many a wild animal there comes at tunes the murder lust which under the spur of opportunity leads to genuine crime. In some of the many cases that have come under my notice, the desire to commit murder for the sake of murder has been as sharply defined as the fangs or horns of the criminal. Of the many emotions of wild animals which are revealed more sharply in captivity than in a state of nature, the crime-producing passions, of jealousy, hatred, desire for revenge, and devilish lust for innocent blood, are most prominent. In the management of large animals in captivity, the criminal instinct is quite as great a trouble- breeder and source of anxiety as are wild-animal diseases, and the constant struggle with the elements.

In many cases there is not the slightest premonitory manifestation of murderous intent on the part of a potential criminal. Indeed, with most cunning wisdom, a wild-animal murderer will often conceal his purpose until outside interference is an impossibility, and the victim is entirely helpless. These manifestations of fiendish cunning and premeditation are very exasperating to those responsible for the care of animals in captivity.

In every well regulated zoological park, solitary confinement is regarded as an unhappy or intolerable condition. Animals that live in herds and groups in large enclosures always exercise more, have better appetites, and are much more contented and happy than individuals that are singly confined.

To visitors, a happy and contented community of deer, antelopes, bears, wolves, or birds is a source of far more mental satisfaction than could be found in any number of solitary animals. A small pen with a solitary animal in it at once suggests the prison-and-prisoner idea, and sometimes arouses pity and compassion rather than pleased admiration. The peaceful herd or flock is the thing to strive for as the highest ideal attainable in an exhibition of wild animals. But mark well the difficulties.

All the obstacles encountered in carrying out the community idea are created by the evil propensities of the animals themselves. Among the hoofed animals generally, every pair of horns and front hoofs is a possible storm-center. No keeper knows whether the members of his herd of deer will live together in peace and contentment until tomorrow, or whether, on any autumn or winter night, a buck will suddenly develop in his antlered head the thought that it is a good time to "kill something."

In the pairing season we always watch for trouble, and the danger signal always is up. In October a male elk may become ever so savage, and finally develop into a raging demon, dangerous to man and beast; but when he first manifests his new temper openly and in the broad light of day, we feel that he is treating fairly both his herd-mates and his keepers. If he gives fair warning to the world about him, we must not class him as a mean criminal, no matter what he may do later on. It is our duty to corral him at night according to the violence of his rage. If we separate him from the herd, and he tears a fence in pieces and kills his rival, that is honest, open warfare, not foul murder. But take the following case.

In October, 1905, the New York Zoological Park received from the state of Washington a young mule deer buck and two does. Being conspicuous members of the worst species of "difficult" deer to keep alive at Atlantic tidewater, and being also very thin and weak, it required the combined efforts of several persons to keep them alive. For six months they moped about their corral, but at last they began to improve. The oldest doe gave birth to two fawns which actually survived. But, even when the next mating season began, the buck continued to be lanquid and blase. At no time did he exhibit signs of temper, of even suspicious vigor.

In the middle of the night of November 6, 1906, without the slightest warning, he decided to commit a murder, and the mother of the two nursing fawns was selected as the victim. Being weak from the rearing of her offspring, she was at his mercy. He gored her most savagely, about twenty times, and killed her.

That was deliberate, fiendish and cowardly murder. The killing of any female animal by her male consort is murder; but there are circumstances wherein the plea of temporary insanity is an admissible defense. In the autumn, male members of the deer family *often become temporarily insane and irresponsible,* and should be judged accordingly. With us, sexual insanity is a recognized disease.

Such distressing cases as the above are so common that whenever I go deer-hunting and kill a lusty buck, the thought occurs to me, — "another undeveloped murderer, perhaps!"

The most exasperating thing about these corral murders is the cunning treachery of the murderers. Here is another typical case: For three years a dainty little male Osceola deer from Florida was as gentle as a fawn and as harmless as a dove. But one crisp morning Keeper Quinn, to whom every doe in his charge is like a foster-daughter, was horrified at finding blood on the absurd little antlers of the Osceola pet. One of the females lay dead in a dark corner where she had been murdered during the night; and this with another and older buck in the same corral which might fairly have been regarded as an offensive rival.

The desire to murder for the sake of killing is born in some carnivorous animals, and by others it is achieved. Among the largest and finest of the felines, the lions and tigers, midnight murders very rarely occur. We never have known one. Individual dislike is shown boldly and openly, and we are given a fair chance to prevent fatalities. Among the lions, tigers, leopards, jaguars and pumas of the New York Zoological Park, there has been but one murder. That was the crime of Lopez, the big jaguar, who richly deserved instant death as a punishment. It was one of the most cunning crimes I have ever seen among wild animals, and is now historic.

For a year Lopez *pretended,* ostentatiously, to be a good-natured animal! Twenty times at least he acted the part of a playful pet, inviting me to reach in and stroke him. At last we decided to give him a cage-mate, and a fine adult female jaguar was purchased. The animals actually tried to caress each other through the bars, and the big male completely deceived us, one and all.

At the end of two days it was considered safe to permit the female jaguar to enter the cage of Lopez. She was just as much deceived as we were. An animal that is afraid always leaves its traveling- cage slowly and unwillingly, or refuses to leave it at all. When the two sets of doors were opened, the female joyously walked into the cage of her treacherous admirer. In an instant, Lopez rushed upon her, seized her whole neck in his powerful jaws, and crushed her cervical vertebrae by his awful bite. We beat him over the head; we spiked him; we even tried to brain him; but he held her, as a bull-dog would hold a cat, until she was dead. He had determined to murder her, but had cunningly concealed his purpose until his victim was fully in his power.

Bears usually fight "on the square," openly and above-board, rarely committing foul murder. If one bear hates another, he attacks at the very first opportunity, He does not cunningly wait to catch the offender at a disadvantage and beyond the possibility of rescue. Sometimes a captive bear kills a cage-mate or mauls a keeper, but not by the sneaking methods of the human assassin who shoots in the dark and runs away.

I do not count the bear as a common criminal, even though at rare intervals he kills a cage-mate smaller and weaker than himself. One killing of that kind, done by Cinnamon Jim to a small black bear that had annoyed him beyond all endurance, was inflicted as a legitimate punishment, and was so recorded. The attack of two large bears, a Syrian and a sloth bear, upon a small Japanese black bear, in which the big pair deliberately attempted to disembowel the small victim, biting him only in the abdomen, always has been a puzzle to me. I cannot fathom the idea which possessed those two ursine minds; but I have no doubt that some of the book-making men who read the minds of wild animals as if they were open books could tell me all about it.

On the ice-pack in front of his stone hut at the north end of the Franz Josef Archipelago Nansen saw an occurrence that was plain murder. A large male polar bear feeding upon a dead walrus was approached across the ice-pack by two polar-bear cubs. The gorging male immediately stopped feeding and rushed toward the small intruders. They turned and fled wildly; but the villain pursued them, far out upon the ice. He overtook them, killed both, and then serenely returned to his solitary feast.

In February, 1907, a tragedy occurred in the Zoological Park which was a close parallel of the Lopez murder. It was a case in which my only crumb of satisfaction was in my ability to say, "I told you so," — than which no consolation can be more barren.

For seven years there had lived together in the great polar bears' den of the Zoological Park two full-grown, very large and fine polar bears. They came from William Hagenbeck's great group, and both were males. Their rough-and-tumble wrestling, both in the swimming pool and out of it, was a sight of almost perennial interest; and while their biting and boxing was of the roughest character, and frequently drew blood, they never got angry, and never had a real fight.

In the autumn of 1906 one of the animals sickened and died, and presently the impression prevailed that the survivor was lonesome. The desirability of introducing a female companion was spoken of, but I was afraid to try the experiment.

By and by, Mr. Carl Hagenbeck, who had handled about forty polar bears to my one, wrote to us, offering a fine female polar as a mate to the survivor. She was conceded to be one-third smaller than the big male, but was fully adult. Without loss of time I answered, declining to make the purchase, on the ground that our male bear would kill the female. It was my belief that even if he did not at once deliberately murder her, he soon would wear her out by his rough play.

Mr. Hagenbeck replied with the assurance that, in his opinion, all would be well; that, instead of a tragedy taking place, the male would be delighted with a female companion, and that the pair would breed. As convincing proof of the sincerity of his views, Mr. Hagenbeck offered to lose half the purchase price of the female bear in the event that my worst fears were realized.

I asked the opinion of our head keeper of bears, and after due reflection he said:

"Why, no; I don't believe he'd kill her. He's not a *bad* bear at all. I think we could work it so that there would be no great trouble."

Mr. Hagenbeck's son also felt sure there would be no tragedy.

Quite against my own judgment of polar-bear character, but in deference to the expert opinion arrayed against mine, I finally yielded. The female bear was purchased, and on her arrival she was placed for three weeks in the large shifting-cage which connects with the eastern side of the great polar bears' den.

The two animals seemed glad to see each other. At once they fraternized through the bars, licked each other's noses, and ate their meals side by side. At night the male always slept as near as possible to his new companion. There was not a sign of ill temper; but, for all that, my doubts were ever present.

At last, after three full weeks of close acquaintance, it was agreed that there was nothing to be gained by longer delay in admitting the female to the large den. But we made preparations for trouble. The door of the sleeping-den was oiled and overhauled and put in thorough working order, so that if the female should dash into it for safety, a keeper could instantly slide the barrier and shut her in. We provided pike-poles, long iron bars, lariats, meat, and long planks a foot wide. Heartily wishing myself a hundred miles away, I summoned all my courage and gave the order:

"Open her door, a foot only, and let her put her head out. Keep him away."

The female bear had not the slightest fear or premonition of danger. Thrusting her head through the narrow opening, she looked upon the world and the open sky above, and found that it was good. She struggled to force the door open wider; and the male stood back, waiting.

"Let her go!" Forcing the door back with her own eager strength, she fearlessly dropped the intervening eighteen inches to the floor of the den, and was free. The very *next second* the male flung his great bulk upon her, and the tragedy was on.

I would not for five thousand dollars see such a thing again. A hundred times in the twenty minutes that followed I bitterly regretted my folly in acting contrary to my own carefully formed conclusions regarding the temper, the strength, and the mental processes of that male bear.

He never left her alone for ten seconds, save when, at five or six different times, we beat him off by literally ramming him away. When she first fell, the slope of the floor brought her near the cage bars, which gave us a chance to fight for her. We beat him over the head; we drove big steel spikes into him; and we rammed him with planks, not caring how many bones we might break. But each time that we beat him off, and the poor harried female rose to her feet, he flung himself upon her anew, crushed her down upon the snow, and fought to reach her throat!

Gallantly the female fought for her life, with six wild men to help her. After a long battle, — it seemed like hours, but I suppose it was between twenty and thirty minutes, the male bear recognized the fact that so long as the female lay near the bars his own punishment would continue and the end would be postponed. Forthwith he seized his victim and dragged her inward and down to the ice that covered the swimming-pool in the centre of the den, beyond our reach. The floor of the den was so slippery from ice and snow that it was utterly unsafe for any of our men to enter and try to approach the now furious animal within striking distance.

Very quickly some choice pieces of fresh meat were thrown within six feet of the bears, in the hope that the male would be tempted away from his victim. In vain! Then, with all possible haste, Keeper Mulvehill coiled a lasso, bravely entered the den, and with the first throw landed the noose neatly around the neck of the male bear. In a second it was jerked taut, the end passed through the bars, and ten eager arms dragged the big bear away from his victim and close up to the bars. Another lariat was put on him to guard against breakages, and no bear ever missed being choked to death by a narrower margin than did that one. That morsel of revenge was sweet. While he was held thus, two men went in and attached a rope to the now dying female, and she was quickly dragged into the shifting-cage.

But the rescue came too late. At the last moment on the ice, the canine teeth of the big bear had severed the jugular vein of the female, and in two minutes after her rescue she was dead. It is my belief that at first the male did not intend to murder the female. I think his first impulse was to play with her, as he had always done with the male comrade of his own size. But the *joy of combat* seized him, and after that his only purpose was to kill. My verdict was, not premeditated murder, but murder in the second degree.

In the order of carnivorous animals, I think the worst criminals are found in the Marten Family (*Mustelidae*); and if there is a more murderous villain than the mink, I have yet to find him out. The mink is a midnight assassin, who loves slaughter for the joy of murder. The wolverine, the marten, mink and weasel are all courageous, savage and merciless. To the wolverine Western trappers accord the evil distinction of being a veritable imp of darkness on four legs. To them he is the arch-fiend, beyond which animal cunning and depravity cannot go. Excepting the profane history of the pickings and stealings of this "mountain devil" as recorded by suffering trappers, I know little of it; but if its instincts are not supremely murderous, its reputation is no index of its character.

The mink, however, is a creature that we know and fear. Along the rocky shores of the Bronx River, even in the Zoological Park, it perversely persisted long after our park-building began. In spite of traps, guns, and poison, and the killing of from three to five annually in our Park, *Putorius vison* would not give up. With us, the only creatures that practiced wholesale and unnecessary murder were minks and dogs. The former killed our birds, and during one awful period when a certain fence was being rebuilt, the latter destroyed several deer. A mink once visited an open-air yard containing twenty-two pinioned laughing gulls, and during that *noche triste* killed all of those ill-fated birds. It did not devour even one, and it sucked the blood of only two or three.

On another tragic occasion a mink slaughtered an entire flock of fifteen gulls; but its joy of killing was short-lived, for it was quickly caught and clubbed to death. A miserable little weasel killed three fine brant geese, purely for the love of murder; and then he departed this life by the powder-and-lead route.

All the year round captive buffalo bulls are given to fighting, and for one bull to injure or kill another is an occurrence all too common. Even in the great twenty-seven thousand acre reserve of the Corbin Blue Mountain Forest Association, fatal fights sometimes occur. It was left to a large bull named Black Beauty, in our Zoological Park herd, to reveal the disagreeable fact that under certain circumstances a buffalo may become a cunning and deliberate assassin.

In the spring of 1904, a new buffalo bull, named Apache, was added to the portion of our herd which up to that time had been dominated by Black Beauty. We expected the usual head-to-head battle for supremacy, succeeded by a period of peace and quiet. It is the law of the herd that after every contest for supremacy the vanquished bull shall accept the situation philosophically, and thereafter keep his place.

At the end of a half-hour of fierce struggle, head-to-head, Black Beauty was overpowered by Apache, and fled from him into the open range. To emphasize his victory, Apache followed him around and around at a quiet walk, for several hours; but the beaten bull always kept a factor of safety of about two hundred feet between himself and the master of the herd. Convinced that Black Beauty would no longer dispute his supremacy, Apache at last pronounced for peace and thought no more about the late unpleasantness. His rival seemed to accept the situation, and rejoined the herd on the subdued status of an ex-president.

For several days nothing occurred; but all the while Black Beauty was biding his time and watching for an opportunity. At last it came. As Apache lay dozing and ruminating on a sunny hill-side, his beaten rival quietly drifted around his resting-place, stealthily secured a good position, and, without a second's warning plunged his sharp horns deep into the lungs of the reclining bull. With the mad energy of pent-up and superheated fury, the assassin delivered stab after stab into the unprotected side of the helpless victim, and before Apache could gain his feet he had been gored many times. He lived only a few minutes.

It was foul murder, fully premeditated; and had Black Beauty been my personal property, he would have been executed for the crime, without any objections, or motions, or appeals, or far-fetched certificates of unreasonable doubt.

During the past twenty years a number of persons have been treacherously murdered by animals they had fed and protected. One of the most deplorable of these tragedies occurred late in 1906, near Montclair, New Jersey. Mr. Herbert Bradley was the victim. While walking through his deer park, he was wantonly attacked by a white-tailed buck and murdered on the spot. At Helena, Montana, a strong man armed with a pitchfork was killed by a bull elk. There have been several other fatalities from elk.

The greater number of such crimes as the above have been committed by members of the Deer Family (deer, elk, moose and caribou). The hollow-horned ruminants seem to be different. I believe that toward their keepers the bison, buffaloes and wild cattle entertain a certain measure of respect that in members of the Deer Family often is totally absent. But there are exceptions; and a very sad and notable case was the murder of Richard W. Rock, of Henry's Lake, Idaho, in 1903.

Dick Rock was a stalwart ranchman in the prime of life, who possessed a great fondness for big-game animals. He lived not far from the western boundry of the Yellowstone Park. He liked to rope elk and moose in winter, and haul them on sleds to his ranch; to catch mountain goats or mule deer for exhibition; and to breed buffaloes. His finest bull buffalo, named Indian, was one of his favorites, and was broken *to ride!* Scores of times Rock rode him around the corral, barebacked and without bridle or halter. Rock felt that he could confidently trust the animal, and he never dreamed of guarding himself against a possible evil day.

But one day the blood lust seized the buffalo, and he decided to assassinate his best friend. The next time Dick Rock entered the corral, closing the gate and fastening it securely,—thus shutting himself in,—the big bull attacked him so suddenly and fiercely that there was not a moment for either escape or rescue. We can easily estimate the suddenness of the attack by the fact that alert and active Dick Rock had not time even to climb upon the fence of the corral, whereby his life would have been saved. With a mighty upward thrust, the treacherous bull drove one of his horns deeply into his master's body, and impaled him so completely and so securely that the man hung there and died there! As a crowning horror, the bull was unable to dislodge his victim, and the body of the ranchman was carried about the corral on the horns of his assassin until the horrified wife went a mile and a half and summoned a neighbor, who brought a rifle and executed the murderer on the spot.

Such sudden onslaughts as this make it unsafe to trust implicitly, and without recourse, to the good temper of any animal having dangerous horns.

If bird-lovers knew the prevalence of the murder instinct among the feathered folk, no doubt they would be greatly shocked. Many an innocent-looking bird is really a natural villain without opportunity to indulge in crime. It is in captivity that the wickedness inherent in wild creatures comes to the surface and becomes visible. In the open, the weak ones manage to avoid danger, and to escape when threatened; but, with twenty birds in one large cage, escape is not always possible. A "happy-family" of a dozen or twenty different species often harbors a criminal in its midst; and when the criminal cunningly waits until all possibilities of rescue are eliminated, an assassination is the result.

[Illustration with caption: RICHARD W. ROCK AND HIS BUFFALO MURDERER This bison treacherously killed the man soon after this picture was made]

[Illustration with caption: "BLACK BEAUTY" MURDERING "APACHE"] Here is a partial list of the crimes in our bird collection during one year:

A green jay killed a blue jay. A jay-thrush and several smaller birds were killed by laughing thrushes, — which simply love to do murder! A nightingale was killed by a catbird and two mockingbirds. Two snake-birds killed a third one — all of them thoroughly depraved villains. Three gulls murdered another; a brown pelican was killed by trumpeter-swans; and a Canada goose was killed by a gull. All these victims were birds in good health.

It is deplorable, but nevertheless true, that in large mixed companies of birds, say where forty or fifty live together, it is a common thing for a sick bird to be set upon and killed, unless rescued by the keepers. In crimes of this class birds often murder their own kind, but they are quite as ready to kill members of other species. In 1902 a sick brant goose was killed by its mates; and so were a red-tailed hawk, two saras cranes, two black vultures, a road-runner, and a great horned owl. An aged and sickly wood ibis was killed by a whooping crane; and a night heron killed its mate.

Strange as it may seem, among reptiles there is far less of real first-degree murder than among mammals and birds. Twenty rattlesnakes may be crowded together in one cage, without a family jar. Even among cobras, perhaps the most irritable and pugnacious of all serpents, I think one snake never wantonly murders another, although about once in twenty years one will try to swallow another. The big pythons and anacondas never fight, nor try to commit murder. And yet, a twenty-foot regal python with a bad heart—like Nansen's polar bear—could easily constrict and kill any available snake of smaller size.

At this moment I do not recall one instance of wanton murder among serpents. It is well known that some snakes devour other snakes; but that is not crime. The record of the crocodilians is not so clear. It is a common thing for the large alligators in our Reptile House to battle for supremacy and in these contests several fatalities have occurred. Some of these occurrences are not of the criminal sort; but when a twelve-foot alligator attacks and kills a six-foot individual, entirely out of his class and far too small to fight with him, it is murder. An alligator will seize the leg of a rival and by violently whirling around on his axis, like a revolving shaft, twist the leg completely off.

Among sea creatures, the clearly defined criminal instinct, as exhibited aside from the never-ending struggle for existence and the quest of food, is rarely observed, possibly because opportunities are so few. The sanguinary exploits of the grampus, or whale-killer, among whales small enough to be killed and eaten, are the onslaughts of a marine glutton in quest of food.

Among the fishes there is one murderer whose evil reputation is well deserved. The common swordfish of the Atlantic, forty miles or so off Block Island or Montauk Point, is not only one of the most fearless of all fishes, but it also is the most dangerous. His fierce attacks upon the boats of men who have harpooned him and seek to kill him are well known, and his unparalleled courage fairly challenges our wonder and admiration. But, unfortunately, the record of the swordfish is stained with crime. When the spirit of murder prompts him to commit a crime in sheer wantonness, he will attack a whale, stab the unfortunate monster again and again, and pursue it until it is dead. This is prompted solely by brutality and murder lust, for the swordfish feeds upon fish, and never attempts to eat any portion of a whale. It can easily be proved that wild animals in a normal state of nature are by no means as much given to murder, either of their own kind or other kinds, as are many races of men. The infrequency of animal murders cannot be due wholly to the many possibilities for the intended victim to escape, nor to difficulty in killing. In every wild species murders are abundantly possible; but it is wholly against the laws of nature for free wild beasts to kill one another in wantonness. It is left to the savage races of men to commit murders without cause, and to destroy one another by fire. The family crimes and cruelties of people both civilized and savage completely eclipse in blackness and in number the doings of even the worst wild beasts. In wild animals and in men, crime is an index to character. The finest species of animals and the noblest races of men are alike distinguished by their abhorrence of the abuse of the helpless and the shedding of innocent blood. The lion, the elephant, the wild horse, the grizzly bear, the orang-utan, the eagle and the whooping crane are singularly free from the criminal instinct. On the other hand, even today Africa contains tribes whose members are actually fond of practicing cruelty and murder. In the Dark Continent there has lived many a "king" beside whom a hungry lion or a grizzly bear is a noble citizen.

XXV

FIGHTING WITH WILD ANIMALS

The study of the intelligence and temperaments of wild animals is by no means a pursuit of academic interest only. Men now are mixing up with dangerous wild beasts far more extensively than ever before, and many times a life or death issue hangs upon the man's understanding of the animal mind. I could cite a long and gruesome list of trainers, keepers and park owners who have been killed by the animals they did not correctly understand.

Not long ago, it was a park owner who was killed by a dangerous deer. Next it was a bull elk who killed the keeper who undertook to show that the animal was afraid of him. In Idaho we saw a death-penalty mistake with a bull buffalo. Recently, in Spain, an American ape trainer was killed by his big male chimpanzee. Recently in Switzerland a snake-charmer was strangled and killed on the stage by her python.

Men who keep or who handle dangerous animals owe it to themselves, their heirs and their assigns to *know the animal mind and temperament, and to keep on the safe side.*

In view of the tragedies and near-tragedies that animal trainers and keepers have been through during the past twenty years, I am desirous of so vividly exhibiting the wild animal mind and temper that at least a few of the mistakes of the past may be avoided in the future. Fortunately I am able to state that thus far no one ever has been killed by an animal in the Zoological Park; but several of our men have been severely hurt. The writer hereof carries two useless fingers on his best hand as a reminder of a fracas with a savage bear. How Dangerous Animals Attack Men. The following may be listed as the wild animals most dangerous to man:

1. In the open: Alaskan brown bears, the grizzly bears, lion, tiger, elephant, leopard, wolf, African buffalo, Indian gaur and buffalo, and gorilla.

All these species are dangerous to the man who meddles with them, either to kill or to capture them. If they are not molested by man, there is very little to fear from any of them save the man-eating lions, and tigers, the northern wolf packs, Alaskan brown bears and rogue elephants.

2. In captivity, or in process of capture: Under this head a special list may be thus composed:

Male elk and deer in the rutting season; male elephants over fifteen years of age; all bears over one year of age, and especially "pet" bears; all gorillas, chimpanzees and orangs over seven years of age (puberty); all adult male baboons, gibbons, rhesus monkeys, callithrix or green monkeys, Japanese red-faced monkeys and large macaques; many adult bison bulls and cows of individually bad temper; also gaur, Old World buffalo, anoa bulls, many individually bad African antelopes, gnus and hartebeests; all lions, tigers, jaguars, leopards, wolves, hyenas, and all male zebras and wild asses over four years of age.

How they attack. The *lion, tiger and bear* launches at a man's head or face a lightning-quick and powerful fore-paw blow that in one stroke tears the skin and flesh in long gashes, and knocks down the victim with stunning force. Before recovery is possible the assailant rushes to the prostrate man and begins to bite or to tear him. Instinctively the fallen man covers his face with his arms, and with the lion, tiger and leopard the arms come in for fearful punishment. It is the way of carnivorous beasts to attack each other head to head and mouth to mouth, and this same instinct leads these animals to focus their initial attacks upon the heads and faces of their human quarry.

After a man-eating lion or tiger has reduced the human victim to a state of non-resistance, the great beast seizes the man by a bite embracing the chest, and with the feet dragging upon the ground rushes off to a place of safety to devour him at leisure. Dr. David Livingston was seized alive by a lion, and carried I forget how many yards without a stop. His left humerus was broken in the onset, but the lion abandoned him without doing him any further serious harm.

Once I could not believe that a lion or a tiger could pick up a man in his mouth and rapidly carry him off, as a fox gets away with a chicken; but when I shot a male tiger weighing 495 pounds, standing 37 inches high and measuring 35 inches around his jaws, I was forever convinced. In the Malay Peninsula Captain Syers told me that a tiger leaped a stockade seven feet high, seized a Chinese woodcutter, leaped out with him, and carried him away.

In a scrimmage with a lion or tiger in the open, the fight is not prolonged. It is a case of kill or be killed quickly. The time of times for steady nerves and perfectly accurate shooting is when a lion, tiger or bear charges the hunter at full speed, beginning sufficiently far away to give the hunter a sporting chance. *The hunter can not afford to be "scared!"* It is liable to cost too much!

The Alaskan brown bear has a peculiar habit. Occasionally he kills the hunter he has struck down, but very often he contents himself with biting his victim on his fleshy parts, *literally from head to foot.* More than one unfortunate amateur hunter has been fearfully bitten without having a bone broken, and without having an important artery or vein severed. Such unfortunates lie upon their faces, with their arms protecting their heads as best they can, and take the awful punishment until the bear tires of it and goes away. Then they *crawl,* on hands and knees, to come within reach of discovery and help. In the annals of Alaska's frontier life there are some heart-rending records of cases such as I have described, coupled with some marvellous recoveries. Strange to say, bear bites or scratches *almost never produce blood poisoning!* This seems very strange, for the bites of lions, tigers and leopards very frequently end in blood poisoning, incurable fever and death. This probably is due to the clean mouth of the omnivorous bear and the infected mouth of the large cats, from putrid meat between their teeth.

The wolf is particularly dangerous to his antagonists, man or beast, from the cutting power of his fearful snap. His molar teeth shear through flesh and small bones like the gash of a butcher's cleaver; and his wide gape and lightning-quick movements render him a very dangerous antagonist. The bite of a wolf is the most dangerous to man of any animal bite to which keepers are liable, and it is the law of zoological gardens and parks that every wolf bite means a quick application of anti- rabies treatment at a Pasteur institute. Personally, I would be no more scared by a wolf-bite than by a feline bite, but the verdict of the jury is,—"it is best to be on the safe side."

Buck elk and deer very, very rarely attack men in the wilds, unless they have been wounded and brought to bay; and then very naturally they fight furiously. It is the attacks of captive or park-bred animals that are most to be feared.

All the deer that I know attack in the same way,—first by a *slow* push forward, in order to come to close quarters *without getting hurt,* and then follows the relentless push, push, push to get up steam for the final raging and death-dealing drive. Even in fighting each other, buck elk and deer do not come together with a long run and a grand crash. Each potential fighter *fears for his own eyes,* and conserves them by a cautious and deliberate engaging process. This is referred to in another chapter.

Fortunately for poor humanity, the same slow and cautious tactics are adopted when a buck deer or wapiti decides to attack a man. This gives the man in the case a chance to put up his defense.

The attacking deer lowers his head, throws his antlers far to the front, and pushes for the body of the man. The instant a tine touches the soft breast or abdomen, he lunges forward to drive it in. But thanks to that life-saving slow start, the man is mercifully afforded a few seconds of time in which to save himself, or at least delay the punishment.

No man ever should enter the enclosure of a "bad" deer, or any buck deer in the rut, without a stout and tough club or pitchfork for defense. Of the two weapons, the former is the best.

In the first place, keep away from all bad deer, especially between October and January first. If you are beset, follow these instructions, as you value your life:

If unarmed, seize the deer by the antlers before he touches your vitals, hold on for all you are worth, and *shout for help. Keep your feet,* just as long as you possibly can. Never mind being threshed about, so long as you keep your feet and keep the tines out of your vitals. Your three hopes are (1) that help will come, (2) or that you can come within reach of a club or some shelter, or (3) that the animal will in some manner decide to desist, — a most forlorn hope.

With a good club, or even a stout walking-stick, you have a fighting chance. As the animal lowers his head and comes close up to impale you on his spears of bone, hit him a smashing blow *across the side of his head, or his nose.* In a desperate situation, *aim at the eye,* and lay on the blows. If your life is in danger from a buck elk or a large deer, do not hesitate about putting out an eye for him. What are a thousand deer eyes compared with a twelve inch horn thrust through your stomach? My standing instructions to our keepers of dangerous animals are: "Save your own life, at all hazards. Don't let a dangerous animal kill you. Kill any animal rather than let it kill you!"

It is useless to strike a charging deer on the top of its head, or on its antlers. Give a sweeping *side* blow for the unprotected cheek and jaw, or the tender nose. There is nothing that a club can do that is so disconcerting as the eye and nose attack, for a badly injured eye always shuts both eyes, automatically. Once when alone in the corral of the axis deer herd, I was treacherously and wantonly attacked by a full-grown buck. I had violated my own rules about going in armed with a stick, and it was lucky for me that the axis deer was not as large as the barasingha or the mule deer. As the buck lowered his head, threw his long, sharp beams straight forward, and pushed for my vitals, I seized him by both antlers, to make my defense. At that he drove forward and nearly upset me. Quickly I let go the right antler and shifted myself to the animal's left side, where by means of the left antler I pulled the struggling buck's head around to my side. Then he began to plunge. Throwing the weight of my chest upon his shoulders I reached over him and with my free hand finally grasped his right foreleg below the knee, and pulled it up clear of the ground. With that I had him.

He tried to struggle free, but I was strong in those days, and angry besides, and he was helpless. Up beside the deer barn, most providentially for the finish, I saw a very beautiful barrel stave. It was the very thing! I worked him over to it, caught it up, and then still holding him by his left antler I laid that stave along his side until he was well punished, and glad when released to rush from that neighborhood.

Female "pet" deer, and female elk, can and do put up dangerous fights with their front hoofs, standing high up on their hind legs and striking fast and furiously. A gentleman of my acquaintance was thus attacked, most unexpectedly, by his pet white-tailed deer doe. She struck about a dozen times for his breast, and his vest and coat were slit open in several places. I once saw two cow elk engage with their front feet in a hot fight, but they did no real damage.

Of course an angry *bison, buffalo or gaur* lowers its head in attacking a man, and seeks to gore and toss him at the same moment. The American bison will start at a distance of ten or twenty yards, and with half lowered head jump forward, grunting "Uh! Uh! Uh!" as he comes. When close up he pauses for a second and poises his head for the toss. That is the man's one chance. At that instant he must strike the animal on the side of his head, and strike hard; and the region of the eye is the spot at which to aim.

Once we were greatly frightened by the determined charge of a savage cow bison upon Keeper McEnroe, who was armed with a short- handled 4-tine pitchfork. As she grunted and came for him we could not refrain from shouting a terrorized warning, "Look out, McEnroe! Look out!"

He looked out. He stood perfectly still, and calmly awaited the onset. The cow rushed close up, and dropped her chin low down for the goring toss. The keeper was ready for her. Swinging his pitchfork he delivered a smashing blow upon the left side of the cow's head, which disconcerted and checked her. Before she could recover herself he smashed her again, and again. Then she turned tail and ran, followed by the shouts of the multitude.

Adult male elephants are among the most dangerous of all wild animals to keep in captivity. They *will* grow bad- tempered with adult age, keepers *will* become careless of danger that is present every day, and a bad elephant often is a cunning and deceitful devil. The strength of an elephant is so great, the toughness of its hide is so pronounced, and the danger of a sudden attack is so permanent that life in a park with a "bad" elephant is one continuous nightmare.

Naturally we have been ambitious to prevent all manner of fatal wild beast attacks upon our keepers. We try our best to provide for their safety, and having done that to the limit we say: "Now it is up to you to preserve your own life. If you can not save yourself from your bad animals, no other person can do it for you!"

Either positively, comparatively or superlatively, a bad elephant is a cunning, treacherous and dangerous animal. We have seen several elephants in various stages of cussedness. Alice, the adult Indian female, is mentally a freak, but she is not vicious save under one peculiar combination of circumstances. Take her outside her yard, and instantly she becomes a storm centre. Gunda was bad to begin with, worse in continuation and murderously worst at his finish. At present Kartoum is dangerous only to inanimate fences and doors.

A wild elephant attacks a hunter by charging furiously and persistently, sometimes making a real man-chase, seizing the man or knocking him down, and then impaling him upon his tusks as he lies. More than one hunter has been knocked down, and escaped the impalement thrust only through the mercy of heaven that caused the tusks to miss him and expend their murderous fury in the ample earth.

On rare occasions an enraged wild elephant deliberately tramples a man to death; and there is one instance on record wherein the elephant held his dead native victim firmly to the ground while he tore him asunder "and actually jerked his arms and legs to some distance."

In captivity a mean elephant kills a keeper, or other person, by suddenly knocking him down, and then either trampling upon him or impaling him.

Gunda, our big male Indian tusker, was the worst elephant with which I ever came in close touch, and we hope never to see his like again. When about ten years old he came to us direct from Assam, and when I saw his big and bulging eyes, and the slits torn in his ears, I recognized him as a bad-tempered animal. I kept my opinions to myself. Two weeks later when we started Gunda's Hindu keeper back toward his native land, I sent for Keepers Gleason and Forester to give them a choice lot of instruction in elephant management. They heard me through attentively, and then Forester said very solemnly:

"Director, I think that is a bad elephant; and I'm afraid of him!"

Keeper Gleason willingly took him over, on condition that he should have sole charge of him, and as long as Gleason remained in our service he managed the elephant successfully. Elsewhere I have spoken at length of Gunda's mind and manners. He went steadily from bad to worse; but we never once really punished him. The time was when there was only one man in the world whom he feared, and would obey, and that was his keeper, Walter Thuman. I have seen that great dangerous beast cower and quake with fear, and back off into a corner, when Thuman's powerful voice yelled at him, and admonished him to behave himself. But all that ended on the day that he "got" Thuman.

On that fateful afternoon, with no visitors present, Thuman opened the outside door, took Gunda by the left ear, and with his steel- shod elephant hook in his left hand started to lead the huge animal out into his yard. Just inside the doorway Gunda thought he saw his chance, and he took it.

With a fierce sidewise thrust of his head he struck his keeper squarely on the shoulder and sent him plunging to the floor in the stall corner nearest him. Then, instantly he wheeled about and started to follow up his attack. In the fall Thuman's hook flew from his hand.

At first Gunda tried to step on him, but he lay so close into the corner that the elephant could not plant his feet so that they would do execution. Then he tried to kneel upon the keeper, with the same result.

Thuman struggled more closely into the corner, and tried hard to pull himself into the refuse box, through its low door; but with his trunk Gunda caught him by a leg and dragged him back. Then he made a fierce downward thrust with his tusks, which were nearly four feet long, to transfix his intended victim.

His left tusk struck the steel-clad wall and shattered into fragments, half way up. The resounding crash of that breaking tusk was what saved Thuman's life.

Gunda thrust again and again with his sound tusk, with the terrified and despairing keeper trying to cling to the broken tusk and save himself. At last the point of the sound tusk drove full and fair through the flat of Thuman's left thigh, as he lay, and stopped against the concrete floor.

Experienced animal men always are listening for sounds of trouble.

In the cage of Alice, three cages and a vestibule distant, Keeper Dick Richards was busily working, when he heard the peculiar crash of that shattered tusk. "What's all that!" said he; and "That's some trouble," was his own answer.

Grabbing his pitchfork he shot out of that cage, ran down the keeper's passage and in about ten seconds' arrived in front of Gunda's cage. And there was Gunda, killing Walter Thuman.

Richards darted in between the widely-separated front bars, gave a wild yell, and with a fierce thrust drove all the tines of his pitchfork into Gunda's unprotected hind-quarters, where the skin was thin and vulnerable.

With a shrill trumpet scream of pain and rage, Gunda whirled away from Thuman, bolted through the door, and rushed madly into his yard.

Keeper Thuman survived, and his recovery was presently accomplished. When I first called to see him he begged me not to kill Gunda for what he had done, or tried to do. In due course Thuman got well, and again took charge of Gunda; but after that the elephant was not afraid of him. We adopted a policy which prevented further accidents, but finally Gunda became a hopeless case of sexual insanity and lust for murder.

When Gunda became most dangerous, we protected our keepers by chaining his feet, and keeping the men out of the reach of his trunk. Because of this, his fury was boundless; and as soon as it was apparent that he was suffering from his confinement and never would be any better, we quickly decided to end it all. He was painlessly put to death, by Mr. Carl E. Akeley, with a single .26 calibre bullet very skilfully sent through the elephant's brain.

Chimpanzees and Orang-Utans attack and fight men just as they attack each other,—by biting the face and neck, and the hands, shoulders and arms. The fighting ape always reaches out, seizes the arm or wrist of the person to be harmed, drags it up to his mouth and bites savagely. As a home illustration of this method of attack, a chimpanzee named Chico in the Central Park Menagerie once bit a finger from the hand of his keeper. In April, 1921, Mr. Ellis Joseph, the animal dealer, was very severely bitten on his face and neck by his own chimpanzee, so much so in fact that eighteen stitches were required to sew up his lacerations.

One excellent thing about the manners of chimpanzees and orang- utans in captivity and on the stage is that they do not turn deadly dangerous all in a moment, as do bears and elephants, and occasionally deer. The ape who is falling from grace goes gradually, and gives warning signs that wise men recognize. They first become strong and boisterous, then they playfully resist and defy the keeper's restraining hand. Next in order they openly become angry at their keepers over trifles, and bristle up, stamp on the floor and savagely yell. It is then that the whip and the stick become not only useless but dangerous to the user, and must be discarded. It is then that new defensive tactics must be inaugurated, and the keeper must see to it that the big and dangerous ape gets no advantage. This means the exercise of good strategy, and very careful management in cage-cleaning. It calls for two cages for each dangerous ape.

There is only one thing in this world of which our three big chimps are thoroughly afraid, and that is an absurd little *toy gun* that cost about fifty cents, and looks it. No matter how bad Boma may be acting, if Keeper Palmer says in a sharp tone, "*Where's that gun!*" Boma hearkens and stops short, and if the "gun" is shown in front of his cage he flies in terror to the top of his second balcony, and cowers in a corner.

Why are those powerful and dangerous apes afraid of that absurd toy? I do not know. Perhaps the answer is—instinct; but if so, how was it acquired? The natives of the chimp country do not have many firearms, and the white man's guns have been seen and heard by not more than one out of every thousand of that chimp population.

Baboons Throw Stones. So far as we are aware, baboons are the only members of the Order Primates who ever deliberately throw missiles as means of offense. In 1922 there was in the New York Zoological Park a savage and aggressive Rhodesian baboon (*Choiropithecus rhodesiae,* Haagner) which throws stones at people whenever he can get hold of such missiles. We have seen him set up against Keeper Palmer and Curator Ditmars a really vigorous bombardment with stones and coal that had been supplied him. His throw was by means of a vigorous underhand pitch, and but for the intervening bars he would have done very good execution.

Keeper Rawlinson, of the Primate House, who was in the Boer War, states that on one occasion when his company was deploying along the steep side of a rock-covered kopje a troop of baboons above them rolled and threw so many stones down at the men that finally two machine guns were let loose on the savage beasts to disperse them.

THE CURTAIN

On one side of the heights above the River of Life stand the men of this little world, — the fully developed, the underdone, and the unbaked, in one struggling, seething mass. On the other side, and on a level but one step lower down, stands the vanguard of the long procession of "Lower" Animals, led by the chimpanzee, the orang and the gorilla. The natural bridge that *almost* spans the chasm lacks only the keystone of the arch.

Give the apes just one thing, — *speech,* — and the bridge is closed!

Take away from a child its sight, speech and hearing, and the whole world is a mystery, which only the hardest toil of science and education ever can reveal. Give back hearing and sight, without speech, and even then the world is only half available. Give a chimpanzee articulate expression and language, and no one could fix a limit to his progress.

Take away from a man the use of one lobe of his brain, and he is rendered speechless.

The great Apes have travelled up the River of Life on the opposite side from Man, but they are only one lap behind him. Let us not deceive ourselves about that. Remember that truth is inexorable in its demands to be heard.

We need not rack our poor, finite minds over the final problem of evolution, or the final destiny of Man and Ape. We cannot prove anything beyond what we see. We do not know, and we never can know, whether the chimpanzee has a "soul" or not; and we cannot *prove* that the soul of man is immortal. If man possesses a soul of lofty stature, why not a soul of lowly stature for the chimpanzee?

We do not know just *where* "heaven" is; and we cannot know until we find it. But what does it all matter on earth, if we keep to the straight path, and rest our faith upon the Great Unseen Power that we call God?

Said the great Poet of Nature in his ode "To a Waterfowl:"

"He who from zone to zone
Guides through the boundless
Sky thy certain flight, In the long way that I must tread
 alone Will lead my steps aright."

CURTAIN.

BY WILLIAM T. HORNADAY

THE MINDS AND MANNERS OF WILD ANIMALS

CAMP FIRES ON DESERT AND LAVA

CAMP FIRES IN THE CANADIAN ROCKIES

TAXIDERMY AND ZOOLOGICAL COLLECTING

TWO YEARS IN THE JUNGLE The Experiences of a Hunter and Naturalist in India, Ceylon, the Malay Peninsula and Borneo. Illustrated. 8 vo.

THE AMERICAN NATURAL HISTORY A Foundation of Useful Knowledge of the Higher Animals of North America. Four Crown Octavo Volumes, Illustrated in colors and half-tones.

THE SAME Royal 8 vo. Complete in one volume.

OUR VANISHING WILD LIFE Its Extermination and Preservation.

www.ingramcontent.com/pod-product-compliance
Lightning Source LLC
Chambersburg PA
CBHW050159230526
45470CB00001B/161